TECHNICAL ENGINEERING AND DESIGN GUIDES AS ADAPTED FROM THE US ARMY CORPS OF ENGINEERS, NO. 16

ROCK FOUNDATIONS

Published by
ASCE Press
American Society of Civil Engineers
345 East 47th Street
New York, New York 10017-2398

ABSTRACT

This manual provides technical criteria and guidance for the design of rock foundations for civil works or other similar large military structures. It provides a minimal standard to be used in planning a satisfactory rock foundation design under
normal conditions. Unusual or special site, loading, or operating conditions may
require advanced analytical designs that are beyond the scope of this material.

Library of Congress Cataloging-in-Publication Data

Rock foundations.
        p.   cm. — (Technical engineering and design guides as adapt
    ed from the U.S. Army Corps of Engineers ; no. 16)Includes bibliographical
    references (p.).
        ISBN 0-7844-0136-5
        1. Foundations. 2. Rock mechanics. 3. Engineering geology.I.
American Society of Civil Engineers. II. United States. Army. Corps of
Engineers. III. Series.
    TA775.R63 1996   95-49328
    624.1'5132—dc20   CIP

        The material presented in this publication has been prepared in
accordance with generally recognized engineering principles and practices, and
is for general information only. This information should not be used without first
securing competent advice with respect to its suitability for any general or specific application.
        The contents of this publication are not intended to be and should not
be construed to be a standard of the American Society of Civil Engineers
(ASCE) and are not intended for use as a reference in purchase specifications,
contracts, regulations, statutes, or any other legal document.
        No reference made in this publication to any specific method, product,
process or service constitutes or implies an endorsement, recommendation, or
warranty thereof by ASCE.
        ASCE makes no representation or warranty of any kind, whether
express or implied, concerning the accuracy, completeness, suitability or utility
of any information, apparatus, product, or process discussed in this publication,
and assumes no liability therefore.
        Anyone utilizing this information assumes all liability arising
from such use, including but not limited to infringement of any patent or patents.

TABLE OF CONTENTS

Chapter 6. Bearing Capacity

Chapter 7. Sliding Stability

Chapter 8. Cut Slope Stability

Chapter 9. Anchorage Systems

Chapter 10. Instrumentation

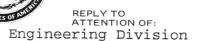

DEPARTMENT OF THE ARMY
U.S. Army Corps of Engineers
WASHINGTON, D.C. 20314-1000

REPLY TO
ATTENTION OF:
Engineering Division

Mr. Charles A. Parthum
President, American Society
    of Civil Engineers
345 East 47th Street
New York, New York 10017

Dear Mr. Parthum:

I am pleased to furnish the American Society of Civil
Engineers (ASCE) a copy of the U. S. Army Corps of Engineers
Engineering Manual, EM 1110-1-2908, Rock Foundations. The Corps
uses this manual to provide technical criteria for design of rock
foundations for large structures.

I understand that ASCE plans to publish this manual for
public distribution. I believe this will benefit the civil
engineering community by improving transfer of technology between
the Corps and other engineering professionals.

Sincerely,

Arthur E. Williams
Lieutenant General, U. S. Army
Commanding

CHAPTER 1

INTRODUCTION

1-1. PURPOSE.

This manual provides technical criteria and guidance for design of rock foundations for civil works or similar large military structures.

1-2. APPLICABILITY.

This manual applies to HQUSACE elements, major subordinate commands, districts, laboratories, and field operating activities.

1-3. REFERENCES.

References pertaining to this manual are listed in Appendix A. References further explain or supplement a subject covered in the body of this manual. The references provided are essential publications to the users of this manual. Each reference is identified in the text by either the designated publication number or by author and date. References to cited material in tables and figures are also identified throughout the manual.

1-4. SCOPE OF MANUAL.

The manual provides a minimum standard to be used for planning a satisfactory rock foundation design for the usual situation. Chapter 2 provides a discussion on design considerations and factor of safety. Chapter 3 provides guidance on site investigation techniques and procedures. Chapter 4 provides guidance on rock mass characterization and classification schemes. Chapters 5 and 6 provide guidance on related topic areas of foundation deformation and settlement and foundation bearing capacity, respectively. Chapters 7 and 8 provide guid-

ance on the sliding stability assessment of gravity structures and slopes cut into rock mass, respectively. Chapter 9 provides guidance on the design of rock anchorage systems. Chapter 10 provides guidance on selection of appropriate geotechnical instrumentation. Chapters 11 and 12 provide discussion on construction considerations and special topics, respectively. Unusual or special site, loading, or operating conditions may warrant sophisticated analytical designs that are beyond the scope of this manual.

1-5. COORDINATION.

A fully coordinated team of geotechnical and structural engineers and engineering geologists should ensure that the result of the analyses are fully integrated into the overall design feature being considered. Some of the critical aspects of the design process that require coordination are the following.

A. Details and Estimates. Exploration details and preliminary estimates of geotechnical parameters, subsurface conditions, and design options.

B. Features. Selection of loading conditions, loading effects, potential failure mechanisms, and other related features of the analytical model.

C. Feasibility. Evaluation of the technical and economic feasibility of alternative structures.

D. Refinement of Design. Refinement of the preliminary design configuration and proportions to reflect consistently the results of more detailed geotechnical site explorations, laboratory testing, and numerical analyses.

E. Unexpected Variations. Modifications to features during construction due to unexpected variations in the foundation conditions.

CHAPTER 2

DESIGN CONSIDERATIONS

2-1. DESIGN APPROACH.

This manual is intended to provide, where possible, a guided approach for the design of rock foundations. The concept of guided design provides for a stepped procedure for solving engineering problems that requires solution by decision making and judgment. Any design that involves rock masses requires a decision-making process in which information must be obtained, considered, and reconciled before decisions and judgments can be made and supported. As such, this manual provides a stepped procedure for planning, collecting, and characterizing the information required to make intelligent decisions and value judgments concerning subsurface conditions, properties, and behavior. A fully coordinated team of geotechnical and structural engineers and engineering geologists are required to ensure that rock foundation conditions and design are properly integrated into the overall design of the structure and that the completed final design of the structure is safe, efficient, and economical. Foundation characterization and design work should be guided by appropriate principles of rock mechanics.

2-2. TYPES OF STRUCTURES.

The types of structures that require analyses as described herein include concrete gravity dams, concrete retaining walls, navigation locks, embankment dams, and similar civil works or military type structures founded on rock. Although directed toward concrete structures, parts of this manual are applicable to all rock foundations.

2-3. DESIGN CONSIDERATIONS.

The design of rock foundations includes two usual analyses, bearing capacity and settlement analyses and sliding stability analyses. Bearing capacity and settlement analyses involve the ability of the rock foundation to support the imposed loads without bearing capacity failure and without excessive or intolerable deformations or settlements. Sliding stability analyses involve the ability of the rock foundation or slope to resist the imposed loads without shearing or sliding. Both analyses must be coordinated and satisfied in a complete design. Basic data that should be obtained during the design stage include strike, dip, thickness, continuity, and composition of all faults and shears in the foundation; depth of overburden; ground water condition; depth of weathering throughout the foundation; joint orientation and continuity; lithology; physical and engineering properties of the rock mass; and loading conditions. Potential failure modes and mechanisms must be determined. For foundation sliding stability, an adequate assessment of the stress conditions and sliding stability of the rock foundation must account for the basic behavior of the structure, the mechanism of transmitting loads to the foundation, the reaction of the foundation to the imposed loads, and the effects of the foundation behavior on the structure. In addition to the above, the analyses of rock foundations must include an evaluation of the effects of seepage and of grouting performed to reduce seepage and the seepage effects. These evaluations are particularly important as related to assessment of hydraulic structures. Because of the difficulty in determining bedrock seepage, seepage paths, and the effectiveness of grouting, conservative assumptions should be used in these evaluations. For a discussion of grouting, see EM 1110-2-3504.

2-4. FACTOR OF SAFETY.

The factor of safety is defined in the manual in terms of the strength parameters of the rock mass. For analyses involving shear or sliding failures, the safety factor is defined as the factor by which the design shear strength must be reduced in order to bring the sliding mass into a state of limiting equilibrium along a given slip plane. This definition pertains to the shear resistance along a given slip surface. The derivation of limit equilibrium equations used to assess sliding stability involves converting stresses to forces. The equations satisfy force equilibrium for the limiting case. For analyses involving bearing capacity failures, the safety factor is defined as the ratio of allowable stress to the actual working stress. The safety factors described in this manual represent the minimum allowable safety factors to be used in the design of rock slopes and foundations for applicable structures. The minimum allowable safety factors described in this manual assume that a complete and comprehensive geotechnical investigation program has been performed. Safety factors greater than the described minimums may be warranted if uncertainties exist in the subsurface conditions or if reliable design parameters cannot be determined. Higher safety factors may also be warranted if un-

usual or extreme loading or operating conditions are imposed on the structure or substructure. Any relaxation of the minimum values involving rock foundations will be subject to the approval of CECW-EG and CECW-ED and should be justified by extensive geotechnical studies of such a nature as to reduce geotechnical uncertainties to a minimum.

CHAPTER 3

SITE INVESTIGATIONS

3-1. SCOPE.

This chapter describes general guidance for site investigation methods and techniques used to obtain information in support of final site evaluation, design, construction, and instrumentation phases of a project with respect to rock foundations. Once a site (preliminary or final) has been selected, the problem usually consists of adapting all phases of the project to existing terrain and rock mass conditions. Because terrain and rock mass conditions are seldom similar between project sites, it is difficult, if not impractical, to establish standardized methodologies for site investigations. In this respect, the scope of investigation should be based on an assessment of geologic structural complexity, imposed or existing loads acting on the foundation, and, to some extent, the consequence should a failure occur. For example, the extent of the investigation could vary from a limited effort where the foundation rock is massive and strong to extensive and detailed where the rock mass is highly fractured and contains weak shear zones. It must be recognized, however, that, even in the former case, a certain minimum of investigation is necessary to determine that weak zones are not present in the foundation. In many cases, the extent of the required field site investigation can be judged from an assessment of preliminary site studies.

3-2. APPLICABLE MANUALS.

Methods and techniques commonly used in site investigations are discussed and described in other design manuals. Two manuals of particular importance are EM 1110-1-1804 and EM 1110-1-1802. It is not the intent of this manual to duplicate material discussed in existing manuals. However, discussions provided in EM 1110-1-1804 and EM 1110-1-1802 apply to both soil and rock. In this respect, this manual will briefly summarize those methods and techniques available for investigating project sites with rock foundations.

SECTION I. PRELIMINARY STUDIES

3-3. GENERAL.

Prior to implementing a detailed site investigation program, certain types of preliminary information will have been developed. The type and extent of information depends on the cost and complexity of the project. The information is developed from a thorough survey of existing information and field reconnaissance. Information on topography, geology and potential geologic hazards, surface and ground-water hydrology, seismology, and rock mass characteristics is reviewed to determine the following:

- Adequacy of available data.
- Type and extent of additional data that will be needed.
- The need for initiating critical long-term studies, such as ground water and seismicity studies, that require advance planning and early action.
- Possible locations and type of geologic features that might control the design of project features.

3-4. MAP STUDIES.

Various types of published maps can provide an excellent source of geologic information to develop the regional geology and geological models of potential or final sites. The types of available maps and their uses are described by Thompson (1979) and summarized in EM 1110-1-1804. EM 1110-1-1804 also provides sources for obtaining published maps.

3-5. OTHER SOURCES OF INFORMATION.

Geotechnical information and data pertinent to the project can frequently be obtained from a careful search of federal, state, or local governments as well as private industry in the vicinity. Consultation with private geotechnical engineering firms, mining companies, well drilling and development companies, and state and private university staff can sometimes provide a wealth of information. EM 1110-1-1804 provides a detailed listing of potential sources of information.

3-6. FIELD RECONNAISSANCE.

After a complete review of available geotechnical data, a geologic field reconnaissance should be made to gather information that can be obtained without subsurface exploration. The primary objective of this initial field reconnaissance is to, insofar as possible, confirm, correct, or expand geologic and hydrologic information collected from preliminary office studies. If rock outcrops are present, the initial field reconnaissance offers an opportunity to collect preliminary information on rock mass conditions that might influence the design and construction of project features. Notation should be made of the strike and dip of major joint sets, joint spacing, joint conditions (i.e., weathering, joint wall roughness, joint tightness, joint infillings, and shear zones), and joint continuity. EM 1110-1-1804 [Murphy (1985)] and Chapter 4 of this manual provide guidance as to special geologic features as well as hydrologic and cultural features that should also be noted.

SECTION II. FIELD INVESTIGATIONS

3-7. GENERAL.

This section briefly discusses those considerations necessary for completion of a successful field investigation program. The majorities of these considerations are discussed in detail in EM 1110-1-1804 and in Chapter 4 of this manual. In this respect, the minimum components that should be considered include geologic mapping, geophysical exploration, borings, exploratory excavations, and in-situ testing. The focus of geologic data to be obtained will evolve as site characteristics are ascertained.

3-8. GEOLOGIC MAPPING.

In general, geologic mapping progresses from the preliminary studies phase with collection of existing maps and information to detailed site-specific construction mapping. Types of maps progress from areal mapping to site mapping to construction (foundation specific) mapping.

A. Areal Mapping. An areal map should consist of sufficient area to include the project site(s) as well as the surrounding area that could

influence or could be influenced by the project. The area and the degree of detail mapped can vary widely depending on the type and size of project and on the geologic conditions. Geologic features and information of importance to rock foundations that are to be mapped include:

- Faults, joints, shear zones, stratigraphy.
- Ground-water levels, springs, surface water, or other evidence of the ground-water regime.
- Potential cavities due to karstic formations, mines, and tunnels.
- Potential problem rocks subject to dissolving, swelling, shrinking, and/or erosion.
- Potential rock slope instability.
- Gas, water, and sewer pipe lines as well as other utilities.

B. Site Mapping. Site maps should be large-scaled with detailed geologic information of specific sites of interest within the project area to include proposed structure areas. A detailed description of the geologic features of existing rock foundation materials and overburden materials is essential in site mapping and subsequent explorations. The determination and description of the subsurface features must involve the coordinated and cooperative efforts of all geotechnical professionals responsible for the project design and construction.

C. Construction Mapping. During construction, it is essential to map the "as-built" geologic foundation conditions as accurately as possible. The final mapping is usually accomplished after the foundation has been cleaned up and just prior to the placement of concrete or backfill. Accurate location of foundation details is necessary. Permanent and easily identified planes of reference should be used. The system of measurement should tie to, or incorporate, any new or existing structure resting on the rock foundation. Foundation mapping should also include a comprehensive photographic record. A foundation map and photographic record will be made for the entire rock foundation and will be incorporated into the foundation report (ER 1110-1-1801). These maps and photographs have proved to be valuable where there were contractor claims, where future modifications to the project became necessary, or where correction of a malfunction or distress of the operational structure requires detailed knowledge of foundation conditions.

3-9. GEOPHYSICAL EXPLORATIONS.

Geophysical techniques consist of making indirect measurements on the ground surface, or in boreholes, to obtain generalized subsurface information. Geologic information is obtained through analysis or interpretation of these measurements. Boreholes or other subsurface explorations are needed for reference and control when geophysical methods are used. Geophysical explorations are of greatest value when performed early in the field exploration program in combination with limited subsurface explorations. The explorations are appropriate for a rapid, though approximate, location and correlation of geologic features such as stratigraphy, lithology, discontinuities, ground water, and for the in-situ measurement of dynamic elastic moduli and rock densities. The cost of geophysical explorations is generally low compared with the cost of core borings or test pits, and considerable savings may be realized by judicious use of these methods. The application, advantages, and limitations of selected geophysical methods are summarized in EM1110-1-1804. EM 1110-1-1802 provides detailed guidance on the use and interpretation of surface and subsurface methods.

3-10. BORINGS.

Borings, in most cases, provide the only viable exploratory tool that directly reveals geologic evidence of the subsurface site conditions. In addition to exploring geologic stratigraphy and structure, borings are necessary to obtain samples for laboratory engineering property tests. Borings are also frequently made for other uses, including collection of ground-water data, performance of in-situ tests, installing instruments, and exploring the condition of existing structures. Boring methods, techniques, and applications are described in EM 1110-1-1804 and EM 1110-2-1907. Of the various boring methods, rock core borings are the most useful in rock foundation investigations.

A. Rock Core Boring. Rock core boring is the process in which diamond or other types of core drill bits are used to drill exploratory holes and retrieve rock core. If properly performed, rock core can provide an almost continuous column of rock that reflects actual rock mass conditions. Good rock core retrieval with a minimum of disturbance requires the expertise of an experienced drill crew.

(1) Standard sizes and notations of diamond core drill bits are summarized in EM 1110-1-1804. Core bits that produce 2.0 in. (nominal) diameter core (i.e., NW or NQ bit sizes) are satisfactory for most exploration work in good rock as well as in providing sufficient size samples for most rock index tests such as unconfined compression, density, and petrographic analysis. However, the use of larger diameter core bits ranging from 4.0 to 6.0 in. (nominal) in diameter are frequently required to produce good core in soft, weak, and/or fractured strata. The larger diameter cores are also more desirable for samples from which rock strength test specimens are prepared, particularly strengths of natural discontinuities.

(2) While the majorities of rock core borings are drilled vertically, inclined borings and, in some cases, oriented cores are required to adequately define stratification and jointing. Inclined borings should be used to investigate steeply inclined jointing in abutments and valley sections for dams, along spillway and tunnel alignments, and in foundations of all structures. In near vertical bedding, inclined borings can be used to reduce the total number of borings needed to obtain core samples of all strata. Where precise geological structure is required from core samples, techniques involving oriented cores are sometimes employed. In these procedures, the core is scribed or engraved with a special drilling tool so that its orientation is preserved. In this manner, both the dip and strike of any joint, bedding plane, or other planar surface can be ascertained.

(3) The number of borings and the depths to which boreholes should be advanced are dependent upon the subsurface geological conditions, the project site areas, types of projects, and structural features. Where rock mass conditions are known to be massive and of excellent quality, the number and depth of boring can be minimal. Where the foundation rock is suspected to be highly variable and weak, such as karstic limestone or sedimentary rock containing weak and compressible seams, one or more boring for each major load bearing foundation element may be required. In cases where structural loads may cause excessive deformation, at least one of the boreholes should be extended to a depth equivalent to an elevation where the structure imposed stress acting within the foundation material is no more than 10% of the maximum stress applied by the foundation. Techniques for estimating structure induced stresses with depth are discussed in Chapter 5 of this manual.

(4) Core logging and appropriate descriptors describing the rock provide a permanent record of the rock mass conditions. Core logging procedures and appropriate rock descriptors are discussed in EM 1110-1-1804, ER 1110-1-1802, Murphy (1985), and Chapter 4 of this manual. Examples of core logs are provided in Appendix D of EM 1110-1-1804. A color photographic record of all core samples should be made in accordance with ER 1110-1-1802.

(5) The sidewalls of the borehole from which the core has been extracted offer a unique picture of the subsurface where all structural features of the rock formation are still in their original position. This view of the rock can be important when portions of rock core have been lost during the drilling operation, particularly weak seam fillers, and when the true dip and strike of the structural features are required. Borehole viewing and photography equipment include borescopes, photographic cameras, TV cameras, sonic imagery loggers, caliper loggers, and alinement survey devices. EP 1110-1-10 provides detailed information on TV and photographic systems, borescope, and televiewer. Sonic imagery and caliper loggers are discussed in detail in EM 1110-1-1802. General discussions of borehole examination techniques are also provided in EM 1110-1-1804.

B. Large-Diameter Borings. Large-diameter borings, 2 ft or more in diameter, are not frequently used. However, their use permits direct examination of the sidewalls of the boring or shaft and provides access for obtaining high-quality undisturbed samples. These advantages are often the principal justification for large-diameter borings. Direct inspection of the sidewalls may reveal details, such as thin weak layers or old shear planes, that may not be detected by continuous undisturbed sampling. Augers are normally used in soils and soft rock, and percussion drills, roller bits, or the calyx method are used in hard rock.

3-11. EXPLORATORY EXCAVATIONS.

Test pits, test trenches, and exploratory tunnels provide access for larger-scaled observations of rock mass character, for determining top of rock profile in highly weathered rock/soil interfaces, and for some in-situ tests that cannot be executed in a smaller borehole.

A. Test Pits and Trenches. In weak or highly fractured rock, test pits and trenches can be constructed quickly and economically by surface-type excavation equipment. Final excavation to grade where samples are to be obtained or in-situ tests performed must be done carefully. Test pits and trenches are generally used only above the groundwater level. Exploratory trench excavations are often used in fault evaluation studies. An extension of a bedrock fault into much younger overburden materials exposed by trenching is usually considered proof of recent fault activity.

B. Exploratory Tunnels. Exploratory tunnels/adits permit detailed examination of the composition and geometry of rock structures such as joints, fractures, faults, shear zones, and solution channels. They are commonly used to explore conditions at the locations of large underground excavations and the foundations and abutments of large dam projects. They are particularly appropriate in defining the extent of marginal strength rock or adverse rock structure suspected from surface mapping and boring information. For major projects where high-intensity loads will be transmitted to foundations or abutments, tunnels/adits afford the only practical means for testing in-place rock at locations and in directions corresponding to the structure loading. The detailed geology of exploratory tunnels, regardless of their purpose, should be mapped carefully. The cost of obtaining an accurate and reliable geologic map of a tunnel is usually insignificant compared with the cost of the tunnel. The geologic information gained from such mapping provides a very useful additional dimension to interpretations of rock structure deduced from other sources. A complete picture of the site geology can be achieved only when the geologic data and interpretations from surface mapping, borings, and pilot tunnels are combined and well correlated. When exploratory tunnels are strategically located, they can often be incorporated into the permanent structure. Exploratory tunnels can be used for drainage and postconstruction observations to determine seepage quantities and to confirm certain design assumptions. On some projects, exploratory tunnels may be used for permanent access or for utility conduits.

3-12. IN-SITU TESTING.

In-situ tests are often the best means for determining the engineering properties of subsurface materials and, in some cases, may be the only way to obtain meaningful results. Table 3-1 lists in-situ tests and their purposes. In-situ rock tests are performed to determine in-situ stresses and deformation properties of the jointed rock mass, shear strength of jointed rock mass or critically weak seams within the rock mass, residual stresses within the rock mass, anchor capacities, and rock mass permeability. Large-scaled in-situ tests tend to average out the effect of complex interactions. In-situ tests in rock are frequently expensive and should be reserved for projects with large, concentrated loads. Well-conducted tests may be useful in reducing overly conservative assumptions. Such tests should be located in the same general area as a proposed structure and test loading should be applied in the

TABLE 3-1. Summary of Purpose and Type of In-Situ Tests for Rock

Purpose of test (1)	Type of test (2)
Strength	Field vane shear[a]
	Direct shear
	Pressuremeter[b]
	Uniaxial compressive[b]
	Borehole jacking[b]
Bearing capacity	Plate bearing[a]
	Standard penetration[a]
Stress conditions	Hydraulic fracturing
	Pressuremeter
	Overcoring
	Flat jack
	Uniaxial (tunnel) jacking[b]
	Chamber (gallery) pressure[b]
Mass deformability	Geophysical (refraction)[c]
	Pressuremeter or dilatometer
	Plate bearing
	Uniaxial (tunnel) jacking[b]
	Borehole jacking[b]
	Chamber (gallery) pressure[b]
Anchor capacity	Anchor/rockbolt loading
Rock mass permeability	Constant head
	Rising or falling head
	Well slug pumping
	Pressure injection

[a]Primarily for clay shales, badly decomposed, or moderately soft rocks, and rock with soft seams.
[b]Less frequently used.
[c]Dynamic deformability.

same direction as the proposed structural loading. In-situ tests are discussed in greater detail in EM 1110-1-1804, the Rock Testing Handbook, and in Chapter 5 of this manual.

SECTION III. LABORATORY TESTING

3-13. GENERAL.

Laboratory tests are usually performed in addition to and after field observations and tests. These tests serve to determine index values for identification and correlation, further refining the geologic model of the site, and they provide values for engineering properties of the rock used in the analysis and design of foundations and cut slopes.

3-14. SELECTION OF SAMPLES AND TESTS.

The selection of samples and the number and type of tests are influenced by local subsurface conditions and the size and type of structure. Prior to any laboratory testing, rock cores should have been visually classified and logged. Selection of samples and the type and number of tests can best be accomplished after development of the geologic model using results of field observations and examination of rock cores, together with other geotechnical data obtained from earlier preliminary investigations. The geologic model, in the form of profiles and sections, will change as the level of testing and the number of tests progresses. Testing requirements are also likely to change as more data become available and are reviewed for project needs. The selection of samples and type of test according to required use of the test results and geological condition is discussed in Chapter 4 of this manual. Additional guidance can be found in EM 1110-2-1902, TM 5-818-1, EM 1110-2-2909, EM 1110-1-1804, Nicholson (1983), Goodman (1976), and Hoek and Bray (1974).

TABLE 3-2. Summary of Purpose and Type of In-Situ Tests for Rock

Purpose of test (1)	Type of test (2)
Strength	Uniaxial compression
	Direct shear
	Triaxial compression
	Direct tension
	Brazilian split
	Point load[a]
Deformability	Uniaxial compression
	Triaxial compress
	Swell
	Creep
Permeability	Gas permeability
Characterization	Water content
	Porosity
	Density (unit weight)
	Specific gravity
	Absorption
	Rebound
	Sonic velocities
	Abrasion resistance

[a]Point load tests are also frequently performed in the field.

3-15. LABORATORY TESTS.

Table 3-2 summarizes laboratory tests according to purpose and type. The tests listed are the types more commonly performed for input to rock foundation analyses and design process. Details and procedures for individual test types are provided in the Rock Testing Handbook. Laboratory rock testing is discussed in Chapter 4 of this manual and in EM 1110-1-1804.

CHAPTER 4

ROCK MASS CHARACTERIZATION

4-1. SCOPE.

This chapter provides guidance in the description and engineering classification of intact rock and rock masses, the types, applications and analyses of rock property tests, the evaluation of intact rock and rock mass properties, and the selection of design parameters for project structures founded on rock. Rock mass characterization refers to the compilation of information and data to build a complete conceptual model of the rock foundation in which all geologic features that might control the stability of project structures, as well as the physical properties of those features, are identified and defined. The compilation of information and data is a continual process. The process starts with the preliminary site investigations and is expanded and refined during site exploration, laboratory and field testing, design analyses, construction, and, in some cases, operation of the project structure. The order of information and data development generally reflects a district's approach to the process but usually evolves from generalized information to the specific details required by the design process. Furthermore, the level of detail required is dependent upon the project structure and the rock mass foundation conditions. For these reasons, this chapter is subdivided into five topic areas according to types of information rather than according to a sequence of tasks. Topic areas include geologic descriptions, engineering classification, shear strength parameters, bearing capacity parameters, and deformation and settlement parameters. The five topic areas provide required input to the analytical design processes described in Chapters 5, 6, 7, and 8.

4-2. INTACT ROCK VERSUS ROCK MASS.

The in-situ rock, or rock mass, is comprised of intact blocks of rock separated by discontinuities such as joints, bedding planes, folds, sheared zones, and faults. These rock blocks may vary from fresh and unaltered rock to badly decomposed and disintegrated rock. Under applied stress, the rock mass behavior is generally governed by the interaction of the intact rock blocks with the discontinuities. For purposes of design analyses, behavioral mechanisms may be assumed as discontinuous (e.g., sliding stability) or continuous (e.g., deformation and settlement).

SECTION I. GEOLOGIC DESCRIPTIONS

4-3. GENERAL.

Geologic descriptions contain some especially important qualitative and quantitative descriptive elements for intact rock and rock masses. Such descriptors are used primarily for geologic classification, correlation of stratigraphic units, and foundation characterization. A detailed description of the foundation rock, its structure, and the condition of its discontinuities can provide valuable insights into potential rock mass behavior. Geologic descriptors can, for convenience of discussion, be divided into two groups: descriptors commonly used to describe rock core obtained during site exploration core boring and supplemental descriptors required for a complete description of the rock mass. Descriptive elements are often tailored to specific geologic conditions of interest. In addition to general geologic descriptors, a number of rock index tests are frequently used to aid in geologic classification and characterization.

4-4. ROCK CORE DESCRIPTORS.

Rock core descriptors refer to the description of apparent characteristics resulting from a visual and physical inspection of rock core. Rock core descriptors are recorded on the drilling log (ENG Form 1836) either graphically or by written description. Descriptions are required for the intact blocks of rock, the rock mass structure (i.e., fractures and bedding), as well as for the condition and type of discontinuity. Criteria for the majorities of these descriptive elements are contained in Table B-2 of EM 1110-1-1804, Table 3-5 of EM 1110-1-1806, and Murphy (1985). Table 4-1 summarizes, consolidates, and, in some instances, expands descriptor criteria contained in the above references. Figures D-6 and D-7 of EM 1110-1-1804 provide examples of typical rock core logs. The following discussions provide a

TABLE 4-1. Summary of Rock Descriptors

1. Intact blocks of rock
    a. Degree of weathering.
        (1) Unweathered: No evidence of any chemical or mechanical alteration.
        (2) Slightly weathered: Slight discoloration on surface; slight alteration along discontinuities; less than 10% of the rock volume altered.
        (3) Moderately weathered: Discoloring evident; surface pitted and altered with alteration penetrating well below rock surfaces; weathering "halos" evident; 10 to 50% of the rock altered.
        (4) Highly weathered: Entire mass discolored; alteration pervading nearly all of the rock with some pockets of slightly weathered rock noticeable; some minerals leached away.
        (5) Decomposed: Rock reduced to a soil with relicit rock texture, generally molded and crumbled by hand.
    b. Hardness.
        (1) Very soft: Can be deformed by hand.
        (2) Soft: Can be scratched with a fingernail.
        (3) Moderately hard: Can be scratched easily with a knife.
        (4) Hard: Can be scratched with difficulty with a knife.
        (5) Very hard: Cannot be scratched with a knife.
    c. Texture.
        (1) Sedimentary rocks:

Texture	Grain diameter	Particle name	Rock name
*	80 mm	Cobble	Conglomerate
*	5–80 mm	Gravel	
Coarse-grained	2–5 mm		
Medium-grained	0.4–2 mm	Sand	Sandstone
Fine-grained	0.1–0.4 mm		
Very fine-grained	0.1 mm	Clay, silt	Shale, claystone, siltstone

   * Use clay-sand texture to describe conglomerate matrix.
        (2) Igneous and metamorphic rocks:

Texture	Grain diameter
Coarse-grained	5 mm
Medium-grained	1–5 mm
Fine-grained	0.1–1 mm
Aphanite	0.1 m

        (3) Textural adjectives: Use simple standard textural adjectives such as prophyritic, vesicular, pegmatitic, granular, and grains well developed, but not sophisticated terms such as holohyaline, hypidimorphic granular, crystal loblastic, and cataclastic.
    d. Lithology macro description of mineral components.
    Use standard adjectives such as shaly, sandy, silty, and calcareous. Note inclusions, concretions, nodules, etc.
2. Rock structure
    a. Thickness of bedding.
        (1) Massive: 3-ft thick or greater.
        (2) Thick-bedded: beds from 1- to 3-ft thick.
        (3) Medium-bedded: beds from 4 in. to 1-ft thick.
        (4) Thin-bedded: 4-in. thick or less.
    b. Degree of fracturing (jointing).
        (1) Unfractured: fracture spacing—6 ft or more.
        (2) Slightly fractured: fracture spacing—2 to 6 ft.
        (3) Moderately fractured: fracture spacing—8 in. to 2 ft.
        (4) Highly fractured: fracture spacing—2 in. to 8 in.
        (5) Intensely fractured: fracture spacing—2 in. or less.

TABLE 4-1. Summary of Rock Descriptors

c. Dip of bed or fracture.
   (1) Flat: 0 to 20 deg.
   (2) Dipping: 20 to 45 deg.
   (3) Steeply dipping: 45 to 90 deg.
3. Discontinuities
  a. Joints.
    (1) Type: Type of joint if it can be readily determined (i.e., bedding, cleavage, foliation, schistosity, or extension).
    (2) Degree of joint wall weathering:
      (i) Unweathered: No visible signs are noted of weathering; joint wall rock is fresh, crystal bright.
      (ii) Slightly weathered joints: Discontinuities are stained or discolored and may contain a thin coating of altered material. Discoloration may extend into the rock from the discontinuity surfaces to a distance of up to 20% of the discontinuity spacing.
      (iii) Moderately weathered joints: Slight discoloration extends from discontinuity planes for greater than 20% of the discontinuity spacing. Discontinuities may contain filling of altered material. Partial opening of grain boundaries may be observed.
      (iv) Highly weathered joints: same as Item 1.a.(4).
      (v) Completely weathered joints: same as Item 1.a.(5).
    (3) Joint wall separations: General description of separation if it can be estimated from rock core; open or closed; if open note magnitude; filled or clean.
    (4) Roughness:
      (i) Very rough: Near vertical ridges occur on the discontinuity surface.
      (ii) Rough: Some ridges are evident; asperities are clearly visible and discontinuity surface feels very abrasive.
      (iii) Slightly rough: Asperities on the discontinuity surface are distinguishable and can be felt.
      (iv) Smooth: Surface appears smooth and feels so to the touch.
      (v) Slickensided: Visual evidence of polishing exists.
    (5) Infilling: Source, type, and thickness of infilling; alterated rock, or by deposition; clay, silt, etc.; how thick is the filler.
  b. Faults and shear zones.
    (1) Extent: Single plane or zone; how thick.
    (2) Character: Crushed rock, gouge, clay infilling, slickensides.

---

brief summary of the engineering significance associated with the more important descriptors.

A. Unit Designation. Unit designation is usually an informal name assigned to a rock unit that does not necessarily have a relationship to stratigraphic rank (e.g., Miami oolite or Chattanooga shale).

B. Rock Type. Rock type refers to the general geologic classification of the rock (e.g., basalt, sandstone, limestone, etc.). Certain physical characteristics are ascribed to a particular rock type with a geological name given according to the rock's mode of origin. Although the rock type is used primarily for identification and correlation, the type is often an important preliminary indicator of rock mass behavior.

C. Degree of Weathering. The engineering properties of a rock can be, and often are, altered to varying degrees by weathering of the rock material. Weathering, which is disintegration and decomposition of the in-situ rock, is generally depth controlled, that is, the degree of weathering decreases with increasing depth below the surface.

D. Hardness. Hardness is a fundamental characteristic used for classification and correlation of geologic units. Hardness is an indicator of intact rock strength and deformability.

E. Texture. The strength of an intact rock is frequently affected, in part, by the individual grains comprising the rock.

F. Structure. Rock structure descriptions describe the frequency of discontinuity spacing and thickness of bedding. Rock mass strength and deformability are both influenced by the degree of fracturing.

G. Condition of Discontinuities. Failure of a rock mass seldom occurs through intact rock but rather along discontinuities. The shear strength along a joint is dependent upon the joint aperture presence or absence of filling materials, the type of the filling material and roughness of the joint surface walls, and pore pressure conditions.

H. Color. The color of a rock type is used not only for identification and correlation, but also for an index of rock properties. Color may be indicative of the mineral constituents of the rock or of the type and degree of weathering that the rock has undergone.

I. Alteration. The rock may undergo alteration by geologic processes at depth, which is distinctively different from the weathering type of alteration near the surface.

4-5. SUPPLEMENTAL DESCRIPTORS.

Descriptors and descriptor criteria discussed in paragraph 4-4 and summarized in Table 4-1 can be readily obtained from observation and inspection of rock core. However, certain important additional descriptors cannot be obtained from core alone. These additional descriptors include orientation of discontinuities, actual thicknesses of discontinuities, first-order roughness of discontinuities, continuity of discontinuities, cavity details, and slake durability.

A. Orientation of Discontinuities. Because discontinuities represent directional planes of weakness, the orientation of the discontinuity is an important consideration in assessing sliding stability and, to some extent, bearing capacity and deformation/settlement. Retrieved core, oriented with respect to vertical and magnetic north, provides a means for determining discontinuity orientation. A number of manufacturers market devices for this purpose. However, most of these techniques abound with practical difficulties (e.g., Hoek and Bray 1974). The sidewalls of the borehole from which conventional core has been extracted offer a unique picture of the subsurface where all structural features of the rock mass are still in their original position. In this respect, techniques that provide images of the borehole sidewalls, such as the borehole camera, the borescope, TV camera, or sonic imagery (discussed in Chapter 3, EM 1110-1-1804, EP 1110-1-10, and EM 1110-1-1802), offer an ideal means of determining the strike and dip angles of discontinuities. The orientation of the discontinuity should be recorded on a borehole photo log. The poles of the planes defined by the strike and dip angles of the discontinuities

should then be plotted on an equal area stereonet. Equal area stereonet pole plots permit a statistical evaluation of discontinuity groupings or sets, thus establishing likely bounds of strike and dip orientations. A stereographic projection plot should then be made of the bounding discontinuity planes for each set of discontinuities to assess those planes that are kinematically free to slide. Goodman (1976), Hoek and Bray (1974), and Priest (1985) offer guidance for stereonet pole plots and stereographic projection techniques.

B. Discontinuity Thickness. The drilling and retrieving of a rock core frequently disturb the discontinuity surfaces. For this reason, aperture measurements of discontinuity surfaces obtained from rock core can be misleading. The best source for joint aperture information is from direct measurement of borehole surface images (e.g., borehole photographs and TV camera recordings). The actual aperture measurement should be recorded on a borehole photo log. An alternative to recording actual measurements is to describe aperture according to the following descriptors:

- Very tight: separations of less than 0.1 mm
- Tight: separations between 0.1 and 0.5 mm
- Moderately open: separations between 0.5 and 2.5 mm
- Open: separations between 2.5 and 10 mm
- Very wide: separations between 10 and 25 mm

For separations greater than 25 mm, the discontinuity should be described as a major discontinuity.

C. First-order Roughness of Discontinuities. First-order roughness refers to the overall, or large scale, asperities along a discontinuity surface. Figure 4-1 illustrates the difference between first-order large-scale asperities and the smaller, second-order asperities commonly associated with

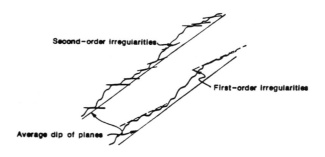

FIG. 4-1. Rough Discontinuity Surface with First-Order and Second-Order Asperities (after Patton and Deere 1970)

roughnesses representative of the rock core scale. The first-order roughness is generally the major contributor to shear strength development along a discontinuity. (See paragraph 4-14b below for further discussion.) A description of this large-scale roughness can only be evaluated from an inspection of exposed discontinuity traces or surfaces. An inspection of rock outcrops in the vicinity of the project site offers an inexpensive means of obtaining this information. Critically oriented joint sets, for which outcrops are not available, may require excavation of inspection adits or trenches. Descriptors such as stepped, undulating, or planar should be used to describe noncritical surfaces. For critically oriented discontinuities, the angles of inclination (referred to as the *i* angle) between the average dip of the discontinuity and first-order asperities should be measured and recorded (Figure 4-1). Hoek and Bray (1974) provide guidance for measuring first-order asperity angles.

D. Continuity of Discontinuities. The continuity of a joint influences the extent to which the intact rock material and the discontinuities separately affect the behavior of the rock mass. In essence, the continuity, or lack of continuity, determines whether the strength that controls the stability of a given structure is representative of a discontinuous rock surface or a combination of discontinuous surfaces and intact rock. For the case of retaining structures, such as gravity dams and lockwalls, a discontinuity is considered fully continuous if its length is greater than the base width in the direction of potential sliding.

E. Cavities. Standard rock coring procedures are capable of detecting the presence of cavities as well as their extent along the borehole axis. However, an evaluation of the volumetric dimensions requires three-dimensional inspection. Downhole TV cameras, with their relatively long focal lengths, provide a means for inspecting cavities. Rock formations particularly susceptible to solutioning (e.g., karstic limestone, gypsum, and anhydrite) may require excavation of inspection trenches or adits to adequately define the location and extent of major cavities. A description of a cavity should include its geometric dimensions, the orientation of any elongated features, and the extent of any infilling, as well as the type of infilling material.

4-6. INDEX TESTS.

Intact samples of rock may be selected for index testing to further aid in geological classification and as indicators of rock mass behavior. As a matter of routine, certain tests will always be performed on representative cores from each major lithological unit and/or weathered class. The number of tests should be sufficient to characterize the range of properties. Routine tests include water content, unit weight, and unconfined compression tests. Additional tests for durability, tensile strength, specific gravity, absorption, pulse velocity, and ultrasonic elastic constants and permeability tests as well as a petrographic examination may be dictated by the nature of the rock or by the project requirements. Types of classification and index tests that are frequently used for rock are listed in Table 4-2.

SECTION II. ROCK MASS CLASSIFICATION

4-7. GENERAL.

Following an appropriate amount of site investigation the rock mass can be divided or classified into zones or masses of similar expected performance. The similar performance may be excavatability, strength, deformability, or any other characteristic of interest, and is determined by use of all of the investigative tools previously described. A good rock mass classification system will:

- Divide a particular rock mass into groups of similar behavior.
- Provide a basis for understanding the characteristics of each group.
- Facilitate planning and design by yielding quantitative data required for the solution of real engineering problems.
- Provide a common basis for effective communication among all persons concerned with a given project.

A meaningful and useful rock mass classification system must be clear and concise, using widely accepted terminology. Only the most significant properties, based on measured parameters that can be derived quickly and inexpensively, should be included. The classification should be general enough that it can be used for a tunnel, slope, or foundation. Because each feature of a rock mass (i.e., discontinuities, intact rock, weathering, etc.) has a different significance, a ranking of combined factors is necessary to satisfactorily describe a rock mass. Each project may need site-specific zoning or rock mass classification, or it may benefit from use of one of the popular existing systems.

TABLE 4-2. Laboratory Classification and Index Tests for Rock

Test (1)	Test method (2)	Remarks (3)
Unconfined (uniaxial) compression	RTH[a] 111	Primary index test for strength and deformability of intact rock; required input to rock mass classification systems.
Point load test	RTH 325	Indirect method to determine unconfined compressive (UC) strength; can be performed in the field on core pieces unsuitable for UC tests.
Water content	RTH 106	Indirect indication of porosity of intact rock or clay content of sedimentary rock.
Unit weight and total porosity	RTH 109	Indirect indication of weathering and soundness.
Splitting strength of rock (Brazilian tensile strength method)	RTH 113	Indirect method to determine the tensile strength of intact rock.
Durability	ASTM[b] D-4644	Index of weatherability of rock exposed in excavations.
Specific gravity of solids	RTH 108	Indirect indication of soundness of rock intended for use as riprap and drainage aggregate.
Pulse velocities and elastic constants	RTH 110	Index of compressional wave velocity and ultrasonic elastic constants for correlation with in-situ geophysical test results.
Rebound number	RTH 105	Index of relative hardness of intact rock cores.
Permeability	RTH 114	Intact rock (no joints or major defects).
Petrographic examination	RTH 102	Performed on representative cores of each significant lithologic unit.
Specific gravity and absorption	RTH 107	Indirect indication of soundness and deformability.

[a]Rock Testing Handbook.
[b]American Society for Testing and Materials.

4-8. AVAILABLE CLASSIFICATION SYSTEMS.

Numerous rock mass classification systems have been developed for universal use. However, six have enjoyed greater use. The six systems include Terzaghi's Rock Load Height Classification (Terzaghi 1946), Lauffer's Classification (Lauffer 1958), Deere's Rock Quality Designation (RQD) (Deere 1964), RSR Concept (Wickham, Tiedemann, and Skinner 1972), Geomechanics System (Bieniawski 1973), and the Q-System (Barton, Lien, and Lunde 1974). Most of the above systems were primarily developed for the design of underground excavations. However, three of the above six classification systems have been used extensively in correlation with parameters applicable to the design of rock foundations. These three classification systems are the Rock Quality Designation, Geomechanics System, and the Q-System.

4-9. ROCK QUALITY DESIGNATION.

Deere (1964) proposed a quantitative index obtained directly from measurements of rock core pieces. This index, referred to as the Rock Quality Designation (RQD), is defined as the ratio (in percent) of the total length of sound core pieces 4 in. (10.16 cm) in length or longer to the length of the core run. The RQD value, then, is a measure of the degree of fracturing, and, since the ratio counts only sound pieces of intact rock, weathering is accounted for indirectly. Deere (1964) proposed the following relationship between the RQD index and the engineering quality of the rock mass.

RQD (%)	Rock Quality
< 25	Very poor
25 < 50	Poor
50 < 75	Fair
75 < 90	Good
90 < 100	Excellent

The determination of RQD during core recovery is simple and straightforward. The RQD index is internationally recognized as an indicator of rock mass conditions and is a necessary input parameter for the Geomechanics System and Q-System. Since core logs should reflect to the maximum extent possible the rock mass conditions encountered, RQD should be determined in the field and recorded on the core logs. Deere and Deere (1989) provides the latest guidance for determining RQD.

4-10. GEOMECHANICS CLASSIFICATION.

A. General. The Geomechanics Classification, or Rock Mass Rating (RMR) system, proposed by Bieniawski (1973), was initially developed for tunnels. In recent years, it has been applied to the preliminary design of rock slopes and foundations as well as for estimating the in-situ modulus of deformation and rock mass strength. The RMR uses six parameters that are readily determined in the field:

- Uniaxial compressive strength of the intact rock
- Rock Quality Designation (RQD)
- Spacing of discontinuities
- Condition of discontinuities
- Ground water conditions
- Orientation of discontinuities

All but the intact rock strength are normally determined in the standard geological investigations and are entered on an input data sheet. (See Table B-1, Appendix B.) The uniaxial compressive strength of rock is determined in accordance with standard laboratory procedures but can be readily estimated on site from the point-load strength index. (See Table 4-2.)

B. Basic RMR Determination. The input data sheet (Table B-1, Appendix B) summarizes, for each core hole, all six input parameters. The first five parameters (i.e., strength, RQD, joint spacing, joint conditions, and ground water) are used to determine the basic RMR. Importance ratings are assigned to each of the five parameters in accordance with Part A of Table B-2, Appendix B. In assigning the rating for each core hole, the average conditions rather than the worst are considered. The importance ratings given for joint spacings apply to rock masses having three sets of joints. Consequently, a conservative assessment is obtained when only two sets of discontinuities are present. The basic rock mass rating is obtained by adding up the five parameters listed in Part A of Table B-2, Appendix B.

C. Adjustment for Discontinuity Orientation. Adjustment of the basic RMR value is required to include the effect of the strike and dip of discontinuities. The adjustment factor (a negative number), and hence the final RMR value, will vary depending upon the engineering application and the orientation of the structure with respect to the orientation of the discontinuities. The adjusted values, summarized in Part B of Table B-2, Appendix B, are divided into five groups according to orientations that range from very favorable to very unfavorable. The determination of the degree of favorability is made by reference to Table B-3 for assessment of discontinuity orientation in relation to dams (Part A) and tunnels (Part B).

D. Rock Mass Class. After the adjustment is made in accordance with Part B, Table B-2, Appendix B, the rock mass ratings are placed in one of five rock mass classes in Part C, Table B-2, Appendix B. Finally, the ratings are grouped in Part D of Table B-2, Appendix B. This section gives the practical meaning of each rock class, and a qualitative description is provided for each of the five rock mass classes. These descriptions range from "very good rock" for class I (RMR range from 81 to 100) to "very poor rock" for class V (RMR ranges < 20). This classification also provides a range of cohesion values and friction angles for the rock mass.

4-11. Q-SYSTEM.

The Q-system, proposed by Barton, Lien, and Lunde (1974), was developed specifically for the design of tunnel support systems. As in the case of the Geomechanics System, the Q-system has been expanded to provide preliminary estimates. Likewise, the Q-system incorporates the following six parameters and the equation for obtaining rock mass quality Q:

- Rock Quality Designation (RQD)
- Number of discontinuity sets
- Roughness of the most unfavorable discontinuity
- Degree of alteration or filling along the weakest discontinuity
- Water inflow
- Stress condition

$$Q = (RQD/J_n) \times (J_r/J_a) \times (J_w/SRF) \quad (4-1)$$

where RQD = Rock Quality Designation; J_n = joint set number; J_r = joint roughness number; J_a = joint alteration number; J_w = joint water reduction number; and SRF = stress reduction number. Table B-4 in Appendix B provides the necessary guidance for assigning values to the six parameters. Depending on the six assigned parameter values reflecting the rock mass quality, Q can vary between 0.001 to 1000. Rock quality is divided into nine classes ranging from exceptionally poor (Q ranging from 0.001 to 0.01) to exceptionally good (Q ranging from 400 to 1000).

4-12. VALUE OF CLASSIFICATION SYSTEMS.

There is perhaps no engineering discipline that relies more heavily on engineering judgment than rock mechanics. This judgment factor is, in part, due to the difficulty in testing specimens of sufficient scale to be representative of rock mass behavior and, in part, due to the natural variability of rock masses. In this respect, the real value of a rock mass classification system is appropriately summarized by Bieniawski (1979). ". . . no matter which classification system is used, the very process of rock mass classification enables the designer to gain a better understanding of the influence of the various geologic parameters in the overall rock mass behavior and, hence, gain a better appreciation of all the factors involved in the engineering problem. This leads to better engineering judgment. Consequently, it does not really matter that there is no general agreement on which rock classification system is best; it is better to try two or more systems and, through a parametric study, obtain a better "feel" for the rock mass. Rock mass classification systems do not replace site investigations, material descriptions, and geologic work-up. They are an adjunct to these items and the universal schemes, in particular, have special value in relating the rock mass in question to engineering parameters based on empirical knowledge."

SECTION III. SHEAR STRENGTH

4-13. GENERAL.

The shear strength that can be developed to resist sliding in a rock foundation or a rock slope is generally controlled by natural planes of discontinuity rather than the intact rock strength. The possible exception to this rule may include structures founded on, or slopes excavated in, weak rock or where a potential failure surface is defined by planes of discontinuities interrupted by segments of intact rock

blocks. Regardless of the mode of potential failure, the selection of shear strength parameters for use in the design process invariably involves the testing of appropriate rock specimens. Selection of the type of test best suited for intact or discontinuous rock, as well as selection of design shear strength parameters, requires an appreciation of rock failure characteristics. Discussions on rock failure characteristics are contained in TR GL-83-13 (Nicholson 1983a) and Goodman (1980).

4-14. ROCK FAILURE CHARACTERISTICS.

Failure of a foundation or slope can occur through the intact rock, along discontinuities or through filling material contained between discontinuities. Each mode of failure is defined by its own failure characteristics.

A. Intact Rock. At stress levels associated with low head gravity dams, retaining walls, and slopes, virtually all rocks behave in a brittle manner at failure. Brittle failure is marked by a rapid increase in applied stress, with small strains, until a peak stress is obtained. Further increases in strain cause a rapid decrease in stress until the residual stress value is reached. While the residual stress value is generally unique for a given rock type and minor principal stress, the peak stress is dependent upon the size of the specimen and the rate that the stress is applied. Failure envelopes developed from plots of shear stress versus normal stress are typically curvilinear.

B. Discontinuities. The typical failure envelope for a clean discontinuous rock is curvilinear, as is intact rock. Surfaces of discontinuous rock are composed of irregularities or asperities ranging in roughness from almost smooth to sharply inclined peaks. Conceptually, there are three modes of failure—asperity override at low normal stresses, failure through asperities at high normal stresses, and a combination of asperity override and failure through asperities at intermediate normal stresses. Typically, those normal stresses imposed by Corps structures are sufficiently low that the mode of failure will be controlled by asperity override. The shear strength that can be developed for the override mode is scale dependent. Initiation of shear displacement causes the override mode to shift from the small-scale second-order irregularities to the large-scale first-order irregularities. As indicated in Figure 4-1, first-order irregularities generally have smaller angles of inclination (i angles) than second-order irregularities. Shear strengths of discontinuities with rough undulating surfaces reflect the largest scale effects with small surface areas (laboratory specimen size) developing higher shear stress than large surface areas (in-situ scale). Figure 4-2 illustrates the influence of both scale effects and discontinuity surface roughnesses.

C. Filled Discontinuities. Failure modes of filled discontinuities can range from those modes associated with clean unfilled discontinuities to those associated with soil. Four factors contribute to their strength behavior: thickness of the filler material, material type, stress history, and displacement history.

(1) Thickness. Research indicates that the strength of discontinuities with filler thicknesses greater than two times the amplitude of the surface undulations is controlled by the strength of the filler material. In general, the thicker the filler material, with respect to the amplitude of the asperities, the less the scale effects.

(2) Material type. The origin of the filler material and the strength characteristics of the joint are important indicators. Sources of filler material include products of weathering or overburden washed into open, water-conducting discontinuities; precipitation of minerals from the ground water; by-products of weathering and alterations along joint walls; crushing of parent rock surfaces due to tectonic and shear displacements; and thin seams deposited during formation. In general, fine-grained clays are more frequently found as fillers and are more troublesome in terms of structural stability.

(3) Stress history. For discontinuities containing fine-grained fillers, the past stress history determines whether the filler behaves as a normally consolidated or overconsolidated soil.

(4) Displacement history. An important consideration in determining the strength of discontinui-

ties filled with fine-grained cohesive materials is whether or not the discontinuity has been subjected to recent displacement. If significant displacement has occurred, it makes little difference whether the material is normally or over-consolidated, since it will be at or near its residual strength.

4-15. FAILURE CRITERIA.

A. Definition of Failure. The term "failure" as applied to shear strength may be described in terms of load, stress, deformation, strain, or other parameters. The failure strengths typically associated with the assessment of sliding stability are generally expressed in terms of peak, residual, ultimate, or as the shear strength at a limiting strain or displacement as illustrated in Figure 4-3. The appropriate definition of failure generally depends on the shape of the shear stress versus shear deformation/strain curve as well as the mode of potential failure. Figure 4-4 illustrates the three general shear stress versus deformation curves commonly associated with rock failure.

B. Linear Criteria. Failure criteria provide an algebraic expression for relating the shear strength at failure with a mathematical model necessary for stability analysis. Mathematical limit equilibrium models used to access sliding stability incorporate the linear Mohr-Coulomb failure criterion as follows:

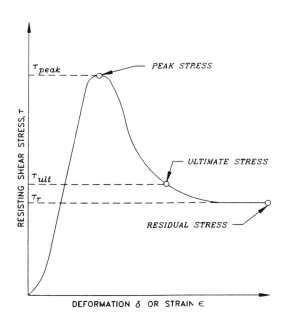

FIG. 4-3. Shear Test Failure as Defined by Peak, Ultimate, and Residual Stress Levels (after Nicholson 1983a)

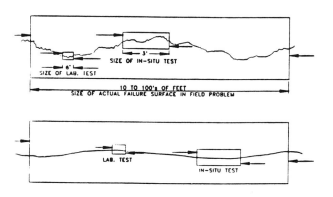

FIG. 4-2. Effect of Different Size Specimens Selected along Rough and Smooth Discontinuity Surface (after Deere et al. 1967)

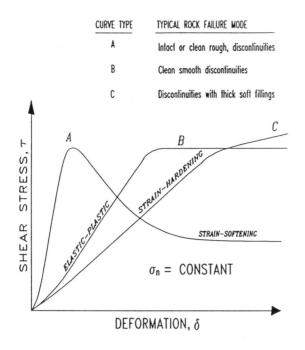

CURVE TYPE	TYPICAL ROCK FAILURE MODE
A	Intact or clean rough, discontinuities
B	Clean smooth discontinuities
C	Discontinuities with thick soft fillings

FIG. 4-4. Hypothetical Shear Stress-Deformation Curves From Drained Direct Shear Tests on (a) Strain-Softening; (b) Elastic-Plastic; and (c) Strain-Hardening Materials (after Nicholson 1983a)

$$\tau_f = c + \sigma_n \tan \phi \qquad (4\text{-}2)$$

where τ_f = the shearing stress developed at failure; and σ_n = stress normal to the failure plane. The c and ϕ parameters are the cohesion intercept and angle of internal friction, respectively. Figure 4-5

illustrates the criterion. It must be recognized that failure envelopes developed from shear tests on rock are generally curved. However, with proper interpretation, failure envelopes over most design stress ranges can be closely approximated by the linear Coulomb equation required by the analytical stability model.

C. Bilinear Criteria. Bilinear criteria (Patton 1966; Goodman 1980) offer a more realistic representation of the shear stress that can be developed along clean (unfilled) discontinuities. These criteria divide a typical curved envelope into two linear segments as illustrated in Figure 4-6. The maximum shear strength that can be developed at failure is approximated by the following equations:

$$\tau_f = \sigma_n \tan (\phi_u + i) \qquad (4\text{-}3)$$

and

$$\tau_f = c_a + \sigma_n \tan \phi_r \qquad (4\text{-}4)$$

where τ_f = maximum (peak) shear strength at failure; σ_n = stress normal to the shear plane (discontinuity); ϕ_u = basic friction angle on smooth planar sliding surface; i = angle of inclination of the first-order (major) asperities; ϕ_r = residual friction angle of the material comprising the asperities; and c_a = apparent cohesion (shear strength intercept) derived from the asperities. For unweathered discontinuity surfaces, the basic friction angle and the residual friction angle are, for practical purposes, the same. The

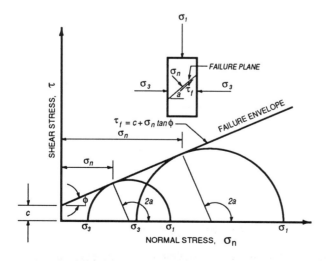

FIG. 4-5. Mohr-Coulomb Relationship with Respect to Principal Stresses and Shear Stress

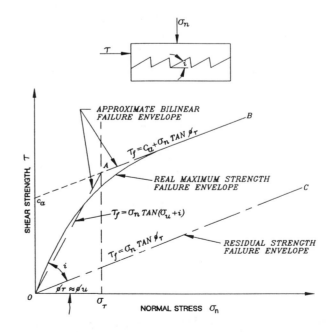

FIG. 4-6. Typical Approximate Bilinear and Real Curvilinear Failure Envelopes for Modeled Discontinuous Rock

intercept of the two equations (i.e., σ_τ in Figure 4-6) occurs at the transition stress between the modes of failure represented by asperity override and shearing of the asperities. Normal stresses imposed by Corps projects are below the transition stress (σ_τ) for the majority of rock conditions encountered. Hence, maximum shear strengths predicted by Equation 4-3 generally control design.

4-16. SHEAR STRENGTH TESTS.

Table 4-3 lists tests that are useful for measuring the shear strength of rock. Details of the tests, test apparatus, and procedures are given in the Rock Testing Handbook (see references Table 4-3), EM 1110-1-1804, and GL-83-14 (Nicholson 1983b.).

4-17. SHEAR STRENGTH TESTING PROGRAM.

The testing program for measuring shear strengths of rock specimens reflects the intended use of the test results (preliminary or final design), the type of specimens (intact or discontinuous), the cost, and, in some cases, the availability of testing devices. In general, the testing program closely parallels the field exploration program, advancing from preliminary testing where modes of potential failure are poorly defined to detailed testing of specific modes of potential failure controlling project design. As a minimum, the following factors should be considered prior to initiating the final detailed phase of testing: the sensitivity of stability with respect to

strengths, loading conditions, suitability of tests used to model modes of failure, and the selection of appropriate test specimens.

A. Sensitivity. A sensitivity analysis should be performed to evaluate the relative sensitivity of the shear strengths required to provide an adequate calculated factor of safety along potential failure planes. Such analysis frequently indicates that conservative and inexpensively obtained strengths often provide an adequate measure of stability, without the extra cost of more precisely defined in-situ strengths.

B. Loading Conditions. Shear tests on rock specimens should duplicate the anticipated range of normal stresses imposed by the project structure along potential failure planes. Duplication of the normal stress range is particularly important for tests on intact rock, or rough natural discontinuities, that exhibit strong curvilinear failure envelopes.

C. Shear Test Versus Mode of Failure. Both triaxial and direct shear tests are capable of providing shear strength results for all potential modes of failure. However, a particular type of test may be considered better suited for modes of failure. The suitability of test types, with respect to modes of failure, should be considered in specifying a testing program.

(1) Laboratory triaxial test. The triaxial compression test is primarily used to measure the undrained shear strength and, in some cases, the elastic properties of intact rock samples subjected to various confining pressures. By orienting planes of weakness, the strength of natural joints, seams, and bedding planes can also be measured. The oriented plane variation is particularly useful for obtaining

TABLE 4-3. Tests to Measure Shear Strength of Rock

Test (1)	Reference (2)	Remarks (3)
Laboratory direct shear	RTH 203[a]	Strength along planes of weakness (bedding), discontinuities or rock-concrete contact; not recommended for intact rock.
Laboratory triaxial	RTH 202	Deformation and strength of inclined compression planes of weakness and discontinuities; strain and strength of intact rock.
In-situ direct shear	RTH 321	Expensive; generally reserved for critically located discontinuities filled with a thin seam of very weak material.
In-situ uniaxial	RTH 324	Expensive; primarily used for defining compression scale effects of weak intact rock; several specimen sizes usually tested.

[a]Rock Testing Handbook.

strength information on thinly filled discontinuities containing soft material. Confining pressures tend to prevent soft fillers from squeezing out of the discontinuity. The primary disadvantage of the triaxial test is that stresses normal to the failure plane cannot be directly controlled. Since clean discontinuities are free draining, tests on clean discontinuities are considered to be drained. Tests on discontinuities filled with fine-grained materials are generally considered to be undrained. (Drained tests are possible but require special testing procedures.) Tests on discontinuities with coarse grained fillers are generally considered to be drained. Detailed procedures for making laboratory triaxial tests are presented in the Rock Testing Handbook (RTH 204).

(2) Laboratory direct shear test. The laboratory direct shear test is primarily used to measure the shear strength, at various normal stresses, along planes of discontinuity or weakness. Although sometimes used to test intact rock, the potential for developing adverse stress concentrations and the effects from shear box induced moments make the direct shear test less than ideally suited for testing intact specimens. Specimen drainage conditions, depending on mode of failure, are essentially the same as for laboratory triaxial tests discussed above. The test is performed on core samples ranging from 2 to 6 in. in diameter. Detailed test procedures are presented in the Rock Testing Handbook (RTH 203).

(3) In-situ direct shear test. In-situ direct shear tests are expensive and are only performed where critically located, thin, weak, continuous seams exist within relatively strong adjacent rock. In such cases, conservative lower bound estimates of shear strength seldom provide adequate assurance against instability. The relatively large surface area tested is an attempt to address unknown scale effects. However, the question of how large a specimen is large enough still remains. The test, as performed on thin, fine-grained, clay seams, is considered to be an undrained test. Test procedure details are provided in the Rock Testing Handbook (RTH 321). Technical Report S-72-12 (Zeigler 1972) provides an indepth review of the in-situ direct shear test.

(4) In-situ uniaxial compression test. In-situ uniaxial compression tests are expensive. The test is used to measure the elastic properties and compressive strength of large volumes of virtually intact rock in an unconfined state of stress. The uniaxial strength obtained is useful in evaluating the effects of scale. However, the test is seldom performed just to evaluate scale effects on strength.

D. Selection of Appropriate Specimens. No other aspect of rock strength testing is more important than the selection of the test specimens that best represents the potential failure surfaces of a foundation. Engineering property tests conducted on appropriate specimens directly influence the analysis and design of projects. As a project progresses, team work between project field personnel and laboratory personnel is crucial in changing type of test, test specimen type, and number of tests when site conditions dictate. The test specimen should be grouped into rock types and subgrouped by unconfined compressive strength, hardness, texture, and structure, or any other distinguishing features of the samples. This process will help in defining a material's physical and mechanical properties. (General guidance on sample selection is provided in EM 1110-1-1804.) However, shear strength is highly dependent upon the mode of failure, i.e., intact rock, clean discontinuous rock, and discontinuities containing fillers. Furthermore, it must be realized that each mode of failure is scale dependent. In this respect, the selection of appropriate test specimens is central to the process of selecting design shear strength parameters.

4-18. SELECTION OF DESIGN SHEAR STRENGTH PARAMETERS.

A. Evaluation Procedures. The rock mass within a particular site is subject to variations in lithology, geologic structure, and the in-situ stress. Regardless of attempts to sample and test specimens with flaws and/or weaknesses present in the rock mass, these attempts, at best, fall short of the goal. The number, orientation, and size relationship of the discontinuities and/or weaknesses may vary considerably, thus affecting load distribution and the final results. In addition to these factors, laboratory results are dependent on the details of the testing procedures, equipment, sampling procedures, and the condition of the sample at the time of the test. The result of these numerous variables is an expected variation in the laboratory test values, which further complicates the problem of data evaluation. The conversion from laboratory-measured strength parameters to in-situ strength parameters requires a careful evaluation and analysis of the geologic and laboratory test data. Also, a combination of experience and judgment is necessary to assess the degree or level of confidence that is required in the selected parameters. As a minimum, the following should be considered: the most likely mode of prototype failure, the factor of safety, the design use, the cost of tests, and the consequence of a failure. A flow diagram illustrating examples of factors to consider in assessing the level of confidence in selected design strengths is shown in Figure 4-7. In general, an

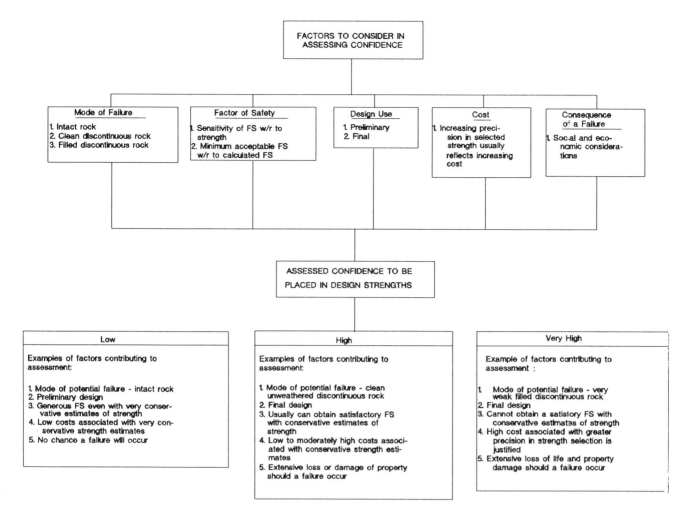

FIG. 4-7. Flow Diagram Illustrating Examples of Factors to Consider in Assessing the Confidence to Be Placed in Selected Design Strengths (after Nicholson 1983a)

increase in assessed confidence should either reflect increasing efforts to more closely define prototype shear strength, at increasing cost, or increasing conservatism in selected design strengths to account for the uncertainties of the in-situ strength.

B. Selection Procedures. Failure envelopes for likely upper and lower bounds of shear strength can generally be determined for the three potential modes of failure; intact rock, clean discontinuities, and filled discontinuities. These limits bound the range within which the in-situ strength is likely to lie. Technical Report GL-83-13 (Nicholson 1983a) describes appropriate test methods and procedures to more accurately estimate in-situ strength parameters. Efforts to more accurately define in-situ strengths must reflect the level of confidence that is required by the design.

(1) Intact rock. Plots of shear stress versus normal stress, from shear test on intact rock, gener-

ally result in considerable data scatter. In this respect, nine or more tests are usually required to define both the upper and lower bounds of shear strength. Figure 4-8 shows a plot of shear stress versus normal stress for a series of tests on a weak limestone. Failure envelopes obtained from a least-squares best fit of upper and lower bounds, as well as all data points, are shown in Figure 4-8. Variations in cohesion values are generally greater than the variations in the friction angle values. With a sufficient number of tests to define scatter trends, over a given range of normal stresses, the confidence that can be placed in the friction angle value exceeds the level of confidence that can be placed in the cohesion value. As a rule, a sufficient factor of safety can be obtained from lower bound estimates of shear strength obtained from laboratory tests. For design cases where lower bound shear strength estimates provide marginal factors of safety, the influ-

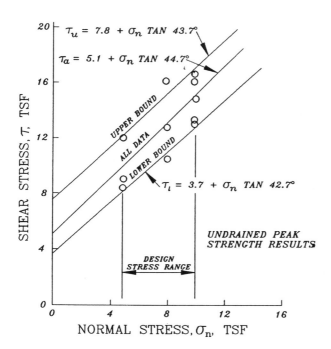

FIG. 4-8. Direct Shear Test Results on Intact Limestone Illustrating Upper and Lower Bounds of Data Scatter

ence of scale effects must be evaluated. Shear strengths obtained from laboratory tests on small specimens should be reduced to account for scale effects. In this respect, Pratt, et al. (1972) and Hoek and Brown (1980) suggest that the full-scale uniaxial compressive strength of intact rock can be as much as 50% lower than the uniaxial compressive strength of a small intact laboratory specimen. In the absence of large scale tests to verify the effects of scale, conservative estimates of the shear strength parameters (cohesion and friction angle), which account for scale effects, can be obtained by reducing the lower bound cohesion value by 50%. This reduced lower bound cohesion value is to be used with the lower bound friction angle value for marginal design cases.

(2) Clean discontinuities. Upper and lower bounds of shear strength for clean discontinuities can be obtained from laboratory tests on specimens containing natural discontinuities and presawn shear surfaces, respectively. The number of tests required to determine the bounds of strength depends upon the extent of data scatter observed in plots of shear stress versus normal stress. As a rule, rough natural discontinuity surfaces will generate more data scatter than smooth discontinuity surfaces. Hence, lower bound strengths obtained from tests on smooth sawn surfaces may require as few as three tests while upper bound strength from tests on very rough natu-

ral engineering judgment cannot be overly emphasized. Discontinuity surfaces may require nine or more tests. Data scatter and/or curvilinear trends in plots of shear stress versus normal stress may result in cohesion intercepts. In such cases, cohesion intercepts are ignored in the selection of design shear strengths. The lower bound failure envelope obtained from shear tests on smooth sawn surfaces defines the basic friction angle (ϕ_u in Equation 4-3). The friction angle selected for design may be obtained from the sum of the basic friction angle and an angle representative of the effective angle of inclination (i in Equation 4-3) for the first-order asperities. The sum of the two angles must not exceed the friction angle obtained from the upper bound shear tests on natural discontinuities. The primary difficulty in selecting design friction values lies in the selection of an appropriate i angle. Discontinuity surfaces or outcrop traces of discontinuities are not frequently available from which to base a reasonable estimate of first-order inclination angles. In such cases, estimates of the i angle must rely on sound engineering judgment and extensive experience in similar geology.

(3) Filled discontinuities. In view of the wide variety of filler materials, previous stress and displacement histories and discontinuity thicknesses, standardization of a procedure for selecting design shear strengths representative of filled discontinuities is difficult. The process is further complicated by the difficulty in retrieving quality specimens that are representative of the discontinuity in question. For these reasons, the use of sound uncertainties associated with unknown conditions effecting shear strength must be reflected in increased conservatism. Generally, the scale effects associated with discontinuous rock are lessened as the filler material becomes thicker in relation to the amplitude of the first-order joint surface undulations. However, potential contributions of the first-order asperities to the shear strength of a filled joint are not considered, as a rule, in the strength selection process because of the difficulty in assessing their effects. Shear strengths that are selected based on in-situ direct shear test of critically located weak discontinuities are the exception to this general rule, but there still remains the problem of appropriate specimen size. As illustrated in Figure 4-9, the displacement history of the discontinuity is of primary concern. If a filled discontinuity has experienced recent displacement, as evident by the presence of slicken-sides, gouge, mismatched joint surfaces, or other features, the strength representative of the joint is at or near its residual value. In such cases, shear strength selection should be based on laboratory residual shear tests of the natu-

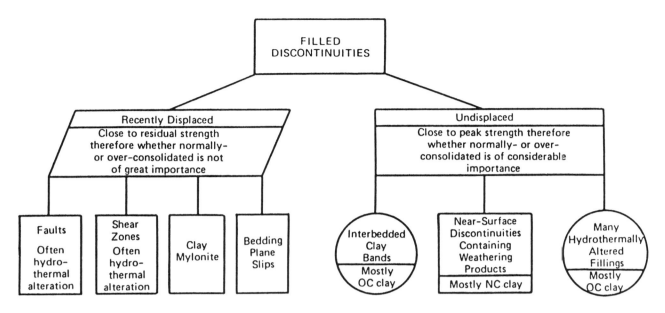

FIG. 4-9. Simplified Division of Filled Discontinuities Into Displaced and Undisplaced and Normally Consolidated (NC) and Overconsolidated (OC) Categories (after Barton 1974)

ral joint. Possible cohesion intercepts observed from the test results should not be included in the selection of design strengths. If the discontinuity has not experienced previous displacement, the shear strength is at or near its peak value. Therefore, whether the filler material is normally or overconsolidated is of considerable importance. In this respect, the shear stress level used to define failure of laboratory test specimens is dependent upon the material properties of the filler. The following definitions of failure stress are offered as general guidance to be tempered with sound engineering judgment: peak strength should be used for filler consisting of normally consolidated cohesive materials and all cohesionless materials; peak or ultimate strength is used for filler consisting of overconsolidated cohesive material of low plasticity; ultimate strength, peak strength of remolded filler, or residual strength is used (depending on material characteristics) for filler consisting of overconsolidated cohesive material of medium to high plasticity.

(4) Combined modes. Combined modes of failure refer to those modes in which the critical failure path is defined by segments of both discontinuous planes and planes passing through intact rock. Selection of appropriate shear strengths for this mode of failure is particularly difficult for two reasons. First, the percentages of the failure path defined by discontinuities or intact rock are seldom known. Second, strains/displacements necessary to cause failure of intact rock are typically an order of magnitude (a factor of 10) smaller than those dis-

placements associated with discontinuous rock. Hence, peak strengths of the intact rock proportion will already have been mobilized and will likely be approaching their residual strength before peak strengths along the discontinuities can be mobilized. For these reasons, selection of appropriate strengths must be based on sound engineering judgment and experience gained from similar projects constructed in similar geological conditions. Shear strength parameters selected for design must reflect the uncertainties associated with rock mass conditions along potential failure paths as well as mechanisms of potential failure (i.e. sliding along discontinuities versus shear through intact rock).

SECTION IV. DEFORMATION AND SETTLEMENT

4-19. GENERAL.

The deformational response of a rock mass is important in seismic analyses of dams and other large structures, as well as the static design of gravity and arch dams, tunnels, and certain military projects. Analytical solutions for deformation and settlement of rock foundations are invariably based on the assumption that the rock mass behaves as a continuum. As such, analytical methods used to compute deformations and the resulting settlements are founded on the theory of elasticity. The selection of design parameters, therefore, involves the selection of appropriate elastic properties: Poisson's ratio and the elastic modulus. Although it is generally recognized that the Poisson's ratio for a rock mass is scale- and stress-dependent, a unique value is frequently assumed. For most rock masses, the Poisson's ratio is between 0.10 and 0.35. As a rule, a poorer quality rock mass has a lower Poisson's ratio than good quality rock. Hence, the Poisson's ratio for a highly fractured rock mass may be assumed as 0.15 while the value for a rock mass with essentially no fractures may be assumed as equal to the value of intact rock. A method for determining Poisson's ratios for intact rock core specimens is described in the Rock Testing Handbook (RTH 201). The selection of an appropriate elastic modulus is the most important parameter in reliable analytical predictions of deformation and settlement. Rock masses seldom behave as an ideal elastic material. Furthermore, modulus is both scale- and stress-dependent. As a result, stress-strain responses typical of a rock mass are not linear. The remaining parts of this section will address appropriate definitions of modulus, scale effects, available methods for estimating modulus values, and the selection of design values.

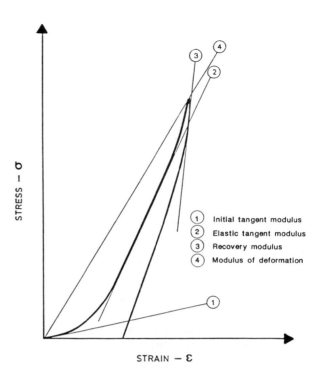

FIG. 4-10. Stress-Strain Curve Typical of In-Situ Rock Mass with Various Moduli that Can Be Obtained

4-20. MODULI DEFINITIONS.

The elastic modulus relates the change in applied stress to the change in the resulting strain. Mathematically, it is expressed as the slope of a given stress-strain response. Since a rock mass seldom behaves as an ideal linear elastic material, the modulus value is dependent upon the proportion of the stress-strain response considered. Figure 4-10 shows a stress-strain curve typical of an in-situ rock mass containing discontinuities with the various moduli that can be obtained. Although the curve, as shown, is representative of a jointed mass, the curve is also typical of intact rock except that the upper part of the curve tends to be concaved downward at stress levels approaching failure. As can be seen in Figure 4-10, there are at least four portions of the stress-strain curve used for determining in-situ rock mass moduli: the initial tangent modulus, the elastic modulus, the tangent recovery modulus, and the modulus of deformation.

A. Initial Tangent Modulus. The initial tangent modulus is determined from the slope of a line constructed tangent to the initial concave upward section of the stress-strain curve (i.e., line 1 in Figure 4-10). The initial curved section reflects the effects of discontinuity closure in in-situ tests and

micro-crack closure in tests on small laboratory specimens.

B. Elastic Modulus. Upon closure of discontinuities/micro-cracks, the stress-strain becomes essentially linear. The elastic modulus, frequently referred to as the modulus of elasticity, is derived from the slope of this linear (or near linear) portion of the curve (i.e., line 2 in Figure 4-10). In some cases, the elastic modulus is derived from the slope of a line constructed tangent to the stress-strain curve at some specified stress level. The stress level is usually specified as 50% of the maximum or peak stress.

C. Recovery Modulus. The recovery modulus is obtained from the slope of a line constructed tangent to the initial segment of the unloading stress-strain curve (i.e., line 3 in Figure 4-10). As such, the recovery modulus is primarily derived from in-situ tests where test specimens are seldom stressed to failure.

D. Modulus of Deformation. Each of the above moduli is confined to specific regions of the stress-strain curve. The modulus of deformation is determined from the slope of the secant line established between zero and some specified stress level (i.e., line 4 in Figure 4-10). The stress level is usually specified as the maximum or peak stress.

4-21. TEST METHODS FOR ESTIMATING MODULUS.

There are at least nine different test methods available to estimate rock modulus. While all nine methods have been used in estimating modulus for design purposes, only the following seven have been standardized: the uniaxial compression tests, uniaxial-jacking tests, the pressuremeter test, plate load test, pressure-chamber tests, radial-jack tests, and borehole-jacking tests. Other test methods that are not standardized but are described in the literature include flat-jack tests and tunnel-relaxation tests.

A. Uniaxial Compression Tests. Laboratory uniaxial compression tests are the most frequently used tests for estimating rock modulus. These tests are performed on relatively small, intact specimens devoid of discontinuities. As such, the results obtained from these tests overestimate the modulus values required for design analyses. Laboratory tests are useful in that the derived moduli provide an upper limit estimate. In-situ uniaxial compression tests are capable of testing specimens of sufficient size to contain a representative number of discontinuities. Modulus values obtained from in-situ tests are considered to be more reliable. This test method is more versatile than some in-situ methods in that test specimens can be developed from any exposed surface. However, the tests are expensive. The Rock Testing Handbook describes test procedures for both laboratory (RTH 201) and in-situ (RTH 324) uniaxial compression tests for the estimation of modulus.

B. Uniaxial Jacking Tests. The uniaxial-jack test involves the controlled loading and unloading of opposing rock surfaces developed in a test adit or trench. The loads are applied by means of large hydraulic jacks that react against two opposing bearing pads. Measurement of the rock mass deformational response below the bearing pads provides two sets of data from which moduli can be derived. The test is expensive. However, the majority of the expense is associated with the excavation of the necessary test adit or trench. The test procedures are described in the Rock Testing Handbook (RTH 365).

C. Pressure Meter Tests. The pressure meter test expands a fluid-filled flexible membrane in a borehole causing the surrounding wall of rock to deform. The fluid pressure and the volume of fluid equivalent to the volume of displaced rock are recorded. From the theory of elasticity, pressure and volume changes are related to the modulus. The primary advantage of the pressure meter is its low cost. The test is restricted to relatively soft rock. Furthermore, the test influences only a relatively small volume of rock. Hence, modulus values derived from the tests are not considered to be representative of rock mass conditions. The test procedures are described in the Rock Testing Handbook (RTH 362).

D. Plate Load Tests. The plate load test is essentially the same as the uniaxial-jacking test except that only one surface is generally monitored for deformation. If sufficient reaction, such as grouted cables, can be provided, the test may be performed on any rock surface. Details of the test procedures are discussed in the Rock Testing Handbook (RTH 364-89).

E. Flat-jack Tests. The flat-jack test is a simple test in which flat-jacks are inserted into a slot cut into a rock surface. Deformation of the rock mass caused by pressurizing the flat-jack is measured by the volumetric change in the jack fluid. The modulus is derived from relationships between jack pressure and deformation. However, analysis of the test results is complicated by boundary conditions imposed by the test configuration. The primary advantages of the test lie in its ability to load a large volume of rock and its relatively low cost. The test procedures are described by Lama and Vutukuri (1978).

F. Pressure-chamber Tests. Pressure-chamber tests are performed in large, underground openings. Generally, these openings are test excavations such as exploratory tunnels or adits. Pre-existing openings, such as caves or mine chambers, can be used if available and applicable to project conditions. The opening is lined with an impermeable membrane and subjected to hydraulic pressure. Instrumented diametrical gauges are used to record changes in tunnel diameter as the pressure load increases. The test is usually performed through several load-unload cycles. The data are subsequently analyzed to develop load-deformation curves from which a modulus can be obtained. The test is capable of loading a large volume of a rock mass from which a representative modulus can be obtained. The test, however, is extremely expensive. The test procedures are described in the Rock Testing Handbook (RTH-361).

G. Radial-Jacking Tests. The radial jacking test is a modification of the pressure chamber test where pressure is applied through a series of jacks placed close to each other. While the jacking system varies, the most common system consists of a series of flat-jacks sandwiched between steel rings and the tunnel walls. The Rock Testing Handbook (RTH-367) describes the test procedures.

H. Borehole-jacking Tests. Instead of applying a uniform pressure to the full cross-section of a borehole as in pressure meter tests, the borehole-jack presses plates against the borehole walls using hydraulic pistons, wedges, or flat-jacks. The technique allows the application of significantly higher pressures required to deform hard rock. The Goodman Jack is the best known device for this test. The test is inexpensive. However, the test influences only a small volume of rock and theoretical problems associated with stress distribution at the plate/rock interface can lead to problems in interpretation of the test results. For these reasons, the borehole-jacking tests are considered to be index tests rather than tests from which design moduli values can be estimated. The tests are described in the Rock Testing Handbook (RTH-368).

I. Tunnel Relaxation Tests. Tunnel relaxation tests involve the measurement of wall rock deformations caused by redistribution of in-situ stresses during tunnel excavation. Except for a few symmetrically shaped openings with known in-situ stresses, back calculations to obtain modulus values from observed deformations generally require numerical modeling using finite element or boundary element computer codes. The high cost of the test is associated with the expense of tunnel excavation.

4-22. OTHER METHODS FOR ESTIMATING MODULUS.

In addition to test methods in which modulus values are derived directly from stress-strain responses of rock, there are at least two additional methods frequently used to obtain modulus values. The two methods include seismic and empirical methods.

A. Seismic Methods. Seismic methods, both downhole and surface, are used on occasion to determine the in-situ modulus of rock. The compressional wave velocity is mathematically combined with the rock's mass density to estimate a dynamic Young's modulus, and the shear wave velocity is similarly used to estimate the dynamic rigidity modulus. However, since rock particle displacement is so small and loading transitory during these seismic tests, the resulting modulus values are nearly always too high. Therefore, the seismic method is generally considered to be an index test. EM 1110-1-1802 and Goodman (1980) describe the test.

B. Empirical Methods. A number of empirical methods have been developed that correlate various rock quality indices or classification systems to in-situ modulus. The more commonly used include correlations between RQD, RMR and Q.

(1) RQD correlations. Deere, Merritt, and Coon (1969) developed an empirical relationship for the in-situ modulus of deformation according to the following formula:

$$E_d = [(0.0231)(RQD) - 1.32] \, E_{t50} \qquad (4\text{-}5)$$

where E_d = in-situ modulus of deformation; RQD = Rock Quality Designation (in percent); and E_{t50} = laboratory tangent modulus at 50% of the unconfined compressive strength. From Equation 4-5 it can be seen that the relationship is invalid for RQD values less than approximately 60%. In addition, the relationship was developed from data that indicated considerable variability between in-situ modulus, RQD, and the laboratory tangent modulus.

(2) RMR correlations. A more recent correlation between in-situ modulus of deformation and the RMR Classification system was developed by Serafim and Pereira (1983) that included an earlier correlation by Bieniawski (1978).

$$E_d = 10^{\frac{RMR - 10}{40}} \qquad (4\text{-}6)$$

where E_d = in-situ modulus of deformation (in GPa); and RMR = Rock Mass Rating value. Equation 4-6 is based on correlations between modulus of deformation values obtained primarily from plate-bearing tests conducted on rock masses of known RMR values ranging from approximately 25 to 85.

(3) Q correlations. Barton (1983) suggested the following relationships between in-situ modulus of deformation and Q values:

$$E_d \text{ (mean)} = 25 \log Q \qquad (4\text{-}7a)$$

$$E_d \text{ (min.)} = 10 \log Q \qquad (4\text{-}7b)$$

$$E_d \text{ (max.)} = 40 \log Q \qquad (4\text{-}7c)$$

where E_d (mean) = mean value of in-situ modulus of deformation (in GPa); E_d(min.) = minimum or lower bound value of in-situ modulus of deformation (in GPa); E_d(max.) = maximum or upper bound value of in-situ modulus of deformation (in GPa); and Q = rock mass quality value.

4-23. CONSIDERATIONS IN SELECTING DESIGN MODULUS VALUES.

Modulus values intended to be representative of in-situ rock mass conditions are subject to extreme variations. There are at least three reasons for these variations: variations in modulus definitions, variability in the methods used to estimate modulus, and rock mass variability.

A. Variations in Modulus Definitions. As noted in paragraph 4-20, the stress-strain responses of rock masses are not linear. Hence, modulus values used in design are dependent upon the portion of the stress-strain curve considered. Because the modulus of deformation incorporates all of the deformation behavior occurring under a given design stress range, it is the most commonly used modulus in analytical solutions for deformation.

B. Variability in Methods. Modulus values obtained from tests are not unique in that the value obtained depends, for the most part, on the test selected. There are at least two reasons for this non-uniqueness. First, with the exception of laboratory compression tests, all of the methods discussed above are in-situ tests in which modulus values are calculated from suitable linear elastic solutions or represent correlations with modulus values derived from in-situ tests. Therefore, the validity of a given method depends to some extent on how well a given solution models a particular test. Finally, the volume of rock influenced by a particular test is a significant factor in how well that test reflects in-situ behavior. Recognizing the potential variation in modulus determinations, the plate-load test has become the most commonly used test for deriving the in-situ modulus of deformation for those projects requiring confidence in estimated values representative of in-situ conditions.

C. Rock Mass Variability. Deformational predictions of foundation materials underlying major project structures such as gravity and arch dams may require analytical solutions for multilayer media. In this respect, the selection of appropriate design deformation moduli will require consideration of not only natural variability within rock layers but also variability between layers.

4-24. SELECTION OF DESIGN MODULI.

As in the selection of design shear strengths, the moduli values used for design purposes are selected rather than determined. The selection proc-

ess requires sound engineering judgment by an experienced team of field and office geotechnical professionals. However, unlike shear strength selection, in which both upper and lower bounds of strength can generally be defined, only the upper bound of the deformation modulus can be readily predicted. This upper bound is derived from unconfined compression tests on intact rock. In addition, the natural variability of the foundation rock, as well as the variability in derived modulus values observed from available methods used to predict modulus, complicates the selection of representative values of modulus. For these reasons, the selection process should not rely on a single method for estimating modulus, but rather the selection process should involve an intergrated approach in which a number of methods are incorporated. Index tests, such as the laboratory unconfined compression test and borehole test devices (Goodman Jack, pressuremeter, and dilatometers), are relatively inexpensive to perform and provide insight as to the natural variability of the rock as well as establish the likely upper bounds of the in-situ modulus of deformation. Empirical correlations between the modulus of deformation and rock mass classification systems (i.e., Equations 4-5, 4-6, and 4-7) are helpful in establishing likely ranges of in-situ modulus values and provide approximate values for preliminary design. Index testing and empirical correlations provide initial estimates of modulus values and form the bases for identifying zones of deformable foundation rock that may adversely effect the performance of project structures. Sensitivity analyses, in which initial estimates of deformation moduli are used to predict deformation response, are essential to define zones critical to design. The design of structures founded on rock judged to be critical to performance must either reflect increasing conservatism in the selected modulus of deformation values or an increase in large scale in-situ testing (i.e., plate-bearing tests, etc.) to more precisely estimate in-situ moduli. The high cost of in-situ tests generally limits the number of tests that can be performed. In this respect, it may not be economically feasible to conduct tests in rock representative of all critical zones, particularly for large projects founded on highly variable rock. In such cases site-specific correlations should be developed between the modulus of deformation values derived from both borehole index tests and large scale in-situ tests and rock mass classification systems (i.e., either the RMR system or the Q-system). If care is taken in selecting test locations, such correlations provide a basis for extrapolating modulus of deformation values that are representative of a wide range of rock mass conditions.

SECTION V. USE OF SELECTED DESIGN PARAMETERS

4-25. GENERAL.

For use of the selected design parameters, refer to the appropriate chapters as follows:

- Chapter 5 - Deformation and Settlement (modulus of deformation).
- Chapter 6 - Bearing Capacity (shear strength).
- Chapter 7 - Sliding Stability (shear strength).
- Chapter 8 - Cut Slope Stability in Rock (shear strength).
- Chapter 9 - Anchorage Systems (shear strength).

CHAPTER 5

DEFORMATION AND SETTLEMENT

5-1. SCOPE.

This chapter describes the necessary elements for estimating and treating settlement, or heave, of structures that are caused by the deformation of the foundation rock. This chapter is subdivided into four sections. Topic areas for the four sections include categories of deformation, analytical methods for predicting the magnitude of deformation, estimating allowable magnitudes of deformation, and methods available for reducing the magnitude of deformation.

SECTION I. CATEGORIES OF ROCK MASS DEFORMATION

5-2. GENERAL.

Deformations that may lead to settlement or heave of structures founded on or in rock may be divided into two general categories: time-dependent deformations and time-independent deformations.

5-3. TIME-DEPENDENT DEFORMATIONS.

Time-dependent deformations can be divided into three different groups according to the mechanistic phenomena causing the deformation. The three groups include consolidation, swelling, and creep.

A. Consolidation. Consolidation refers to the expulsion of pore fluids from voids due to an increase in stress. As a rule, consolidation is associated with soils rather than rock masses. However, rock masses may contain fractures, shear zones, and seams filled with clay or other compressible soils. Sedimentary deposits with interbedded argillaceous rock such as shales and mud stones may also be susceptible to consolidation if subjected to sufficiently high stresses. Consolidation theory and analytical methods for predicting the magnitude of consolidation are addressed in EM 1110-1-1904 and in Instruction Report K-84-7 (Templeton 1984).

B. Swelling. Certain expansive minerals, such as montmorillonite and anhydrite, react and swell in contact with water. Upon drying, these minerals are also susceptible to shrinking. The montmorillonite minerals are generally derived from alteration of ferromagnesian minerals, calcic feldspars, and volcanic rocks and are common in soils and sedimentary rocks. Anhydrite represents gypsum without its water of crystallization and is usually found as beds or seams in sedimentary rock as well as in close association with gypsum and halite in the evaporite rocks. Guidance on procedures and techniques for predicting the behavior of foundations on or in swelling minerals is contained in EM 1110-1-1904, TM 5-818-1, and Miscellaneous Paper GL-89-27 (Johnson 1989).

C. Creep. Creep refers to a process in which a rock mass continues to strain with time upon application of stress. Creep can be attributed to two different mechanisms: mass flow and propagation of microfractures. Mass flow behavior is commonly associated with certain evaporite rock types such as halite and potash. Creep associated with microfracture propagation has been observed in most rock types. Figure 5-1 shows a typical strain-time curve for various constant stress levels. As indicated in Figure 5-1, the shapes of the strain-time curve are a function of the magnitude of the applied stress. Creep will generally occur if the applied stress is within the range associated with nonstable fracture propagation. The transition between stable and nonstable fracture propaga-

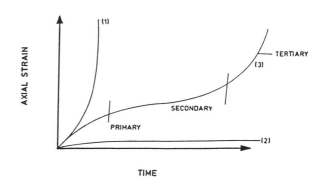

FIG. 5-1. Postulated Strain-Time Curves at (1) Very High Maintained Stress Levels; (2) Moderate Maintained Stress Levels; and (3) High Maintained Stress Levels (from Farmer 1983)

tion varies, depending upon rock type, but typically is on the order of, at least, 50% of the uniaxial compressive strength. Most structures founded on rock generate stress levels well below the transition level. Hence, creep is generally not a problem for the majority of Corps projects. Structures founded on weak rock are the possible exceptions to this rule. Although standardized procedures are available to estimate creep properties of intact rock specimens (i.e., RTH-205), what these properties mean in terms of rock mass behavior is poorly understood. For this reason, estimates of creep response for structures founded on rock masses require specialized studies and, in some cases, research.

5-4. TIME-INDEPENDENT DEFORMATIONS.

Time-independent deformations refer to those deformations that are mechanistically independent of time. Time-independent deformations include deformations generated by prefailure elastic strains, post-failure plastic strains, and deformations resulting from large shear-induced or rotational displacements. Prudent foundation designs preclude consideration of post-failure behavior. Hence, time-independent deformations, as relating to foundation design, refer to deformations that occur as a result of prefailure elastic strains. Analytical methods for estimating rock mass deformations discussed in Section II of this chapter pertain to elastic solutions.

SECTION II. ANALYTICAL METHODS

5-5. GENERAL.

Analytical methods for calculating deformations of foundations may be divided into two general groups: closed form mathematical models and numerical models. The choice of a method in design use depends on how well a particular method models the design problem, the availability, extent, and precision of geological and structural input parameters, the intended use of calculated deformations (i.e., preliminary or final design), and the required accuracy of the calculated values.

5-6. CLOSED-FORM METHODS.

Closed-form methods refer to explicit mathematical equations developed from the theory of elasticity. These equations are used to solve for stresses and strains/deformations within the foundation rock as a function of structure geometry, load and rigidity, and the elastic properties of the foundation rock. Necessary simplifying assumptions associated with the theory of elasticity impose certain limitations on the applicability of these solutions. The most restrictive of these assumptions is that the rock is assumed to be homogeneous, isotropic, and linearly elastic. Poulos and Davis (1974) provide a comprehensive listing of equations, tables, and charts to solve for stresses and displacements in soils and rock. Complex loadings and foundation shapes are handled by superposition in which complex loads or shapes are reduced to a series of simple loads and shapes. Conditions of anisotropy, stratification, and inhomogeneity are treated with conditional assumptions. If sound engineering judgment is exercised to ensure that restrictive and conditional assumptions do not violate reasonable approximations of prototype conditions, closed-form solutions offer reasonable predictions of performance.

A. Input Parameters. Closed-form solutions require, as input parameters, the modulus of elasticity and Poisson's ratio. For estimates of deformation/settlement in rock, the modulus of deformation, E_d, is used in place of modulus of elasticity. Techniques for estimating the modulus of deformation are described in Chapter 4 of this manual. Poisson's ratio typically varies over a small range from 0.1 to 0.35. Generally, the ratio values decrease with decreasing rock mass quality. Because of the small range of likely values and because solutions for deformation are relatively insensitive to assigned values, Poisson's ratio is usually assumed.

B. Depth of Influence. Stresses within the foundation rock that are a result of foundation loads decrease with depth. In cases where the foundation is underlain by multi-layered rock masses, with each layer having different elastic properties, the depth of influence of the structural load must be considered. For the purpose of computing deformation/settlement, the depth of influence is defined as the depth at which the imposed stress acting normal to the foundation plane diminishes to 20% of the maximum stress applied by the foundation. If there is no distinct change in the elastic properties of the subsurface strata within this depth, elastic solutions for layered media need not be considered. Poulos and Davis (1974) and Naval Facilities Engineering Command, NAVFAC DM-7.1 (1982) provide equations and charts based on Boussinesq's equations for estimating stresses with depth imposed by various foundation shapes and loading conditions.

C. Layered Foundation Strata. Poulos and Davis (1974) provided procedures for estimating the deformation/settlement of foundations with the depth of influence for up to four different geologic layers. Multi-layer strata, in which the ratios of moduli of deformation of any of the layers does not exceed a factor of three, may be treated as a single layer with a representative modulus of deformation equivalent to the weighted average of all layers within the depth of influence. A weighted average considers that layers closer to the foundation influence the total deformation to a greater extent than deeper layers. Figure 5-2 shows a foundation underlain by a multi-layer strata containing n number of layers within the depth of influence. The weighted average modulus of deformation may be obtained from Equation 5-1.

$$E_{dw} = \frac{\sum\limits_{i=1}^{n}\left(E_i / \sum\limits_{j=1}^{i} h_j\right)}{\sum\limits_{i=1}^{n}\left(1 / \sum\limits_{j=1}^{i} h_j\right)} \qquad (5\text{-}1)$$

where E_{dw} = weighted average modulus of deformation; $E_{di}, E_{di+1} \ldots E_{dn}$ = modulus of deformation of each layer. The ratios of any $E_{di}, E_{di+1} \ldots E_{dn}$ terms <3; $h_i, h_{i+1} \ldots h_n$ = thickness of each layer; and n = number of layers.

D. Solutions for Uniformly Loaded Rectangular Foundations.
Rectangular founda-

tions are common shapes for footings and other structures. Solutions for deformation of uniformly loaded foundations are divided into two categories: flexible foundations and rigid foundations.

(1) Flexible foundations. Flexible foundations lack sufficient rigidity to resist flexure under load. As indicated in Figure 5-3, the maximum deformation of a uniformly loaded flexible rectangular foundation occurs at the center of the foundation. The maximum deformation (point a in Figure 5-3) can be estimated from the solution of Equation 5-2.

$$\delta_a = \frac{1.12\ qB\ (1 - \mu^2)\ (L/B)^{1/2}}{E_d} \qquad (5\text{-}2)$$

where δ_a = maximum deformation (deformation at point a in Figure 5-3); q = unit load (force/area); B = foundation width; L = foundation length; μ = Poisson's ratio of the foundation rock; and E_d = modulus of deformation of the foundation rock. Estimates of the deformation of points b, c, and d in Figure 5-3 can be obtained by multiplying the estimated deformation at point a (Equation 5-2) by a reduction factor obtained from Figure 5-4.

(2) Rigid foundations. Rigid foundations are assumed to be sufficiently rigid to resist flexure under load. Examples include concrete gravity structures such as intake and outlet structures. Rigid uniformly

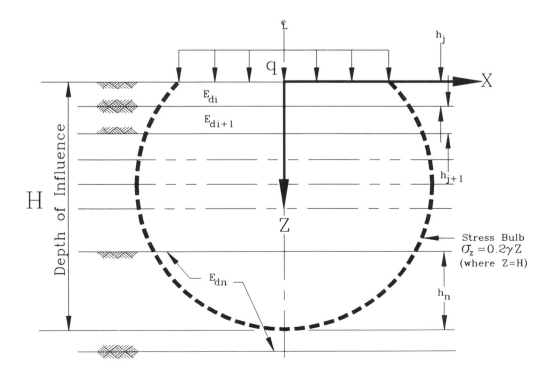

FIG. 5-2. Hypothetical Foundation Underlain by Multilayer Strata Containing _n_ Number of Layers Within Depth of Influence

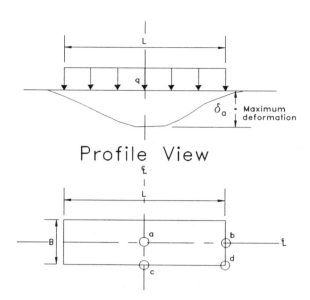

FIG. 5-3. Typical Deformation Profile Under Uniformly Loaded, Rectangular-Shaped, Flexible Foundation

loaded foundations settle uniformly. The estimated deformation can be obtained by multiplying the maximum estimated deformation for a flexible foundation of the same dimensions from Equation 5-2 by the reduction factor obtained from the average for rigid load curve in Figure 5-4.

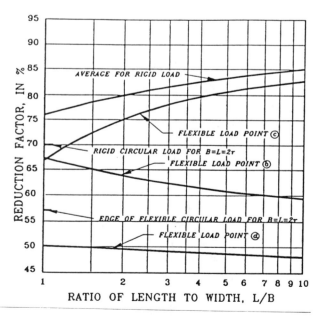

FIG. 5-4. Reduction Factor in Percent of Settlement under Center of Flexible Rectangular-Shaped Foundation (from NAVDOCKS DM-7)

E. Linearly Varying Loads. In practice, most gravity-retaining structures, such as monoliths of gravity dams and lock walls, do not uniformly distribute loads to the foundation rock. As indicated in Figure 5-5, loading of these structures may be approximated by assuming linearly varying load distributions. A complete deformation/settlement analysis requires the calculation of deformations in both the horizontal and vertical planes. Closed-form solutions are available to address linearly varying loads (Poulos and Davis 1974). However, a complete solution requires that the loading conditions be divided into a number of segments. The calculated deformations of each segment are summed to provide a complete solution. In this respect, closed-form solutions are tedious, and, because of simplifying assumptions, provide only approximate solutions.

5-7. NUMERICAL MODELS.

Numerical models refer to those analytical methods that, because of their complexity, require the solution of a large number of simultaneous equations. Such solutions are only reasonably possible with the aid of a computer. In many cases, numerical models provide the only practical alternative for estimating deformation/settlement of structures subjected to complicated loading conditions and/or are founded on anisotropic, nonhomogeneous rock. Numerical approaches can be separated into two general groups: discontinuum and continuum.

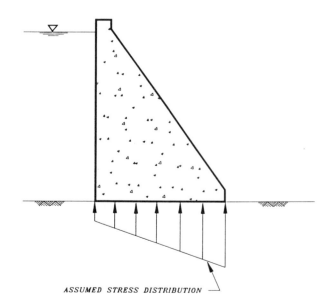

FIG. 5-5. Assumed Linearly Varying Stress Distribution

A. Discontinuum Models. Discontinuum models feature numerical approaches involving equations of motion for rigid particles or blocks. Such models are frequently referred to as discrete element models. Discontinuum approaches are primarily used when analyzing the stability and/or kinematics of one or more independent and recognizable rock blocks. Because the rock blocks are treated as rigid bodies, discontinuum models are not used to analyze magnitudes of rock deformations.

B. Continuum Models. Continuum approaches include the finite element, finite difference, and boundary element methods. All these methods may be used to solve for estimated magnitudes of deformation/settlement. However, the finite element method is the most popular. Numerical modeling of foundation responses dictates the use of constitutive relationships which define material stress-strain behavior. Finite element codes are available that incorporate sophisticated constitutive relationships capable of modeling a variety of nonlinear and/or time-dependent stress-strain behavior. Analytical capabilities offered by some of the more sophisticated codes exceed the ability of the geotechnical engineer to provide meaningful material property parameters. For foundation stress levels and underlying rock types encountered for the majority of structures, reasonable estimates of deformation/settlement can be obtained from linear elastic codes with the modulus of deformation as the primary input parameter. Table 5-1, although not all inclusive, summarizes some of the finite element codes that are commercially available. The choice of code to use should reflect the ability of the code to model the problem at hand and the preference of District office geotechnical professionals charged with the responsibility of settlement analyses.

SECTION III. ALLOWABLE SETTLEMENT

5-8. GENERAL.

For structures founded on rock, the total deformation/settlement seldom controls design. The design for, or control of, differential settlement between critical elements of a structure is essential for the proper and safe functioning of that structure. The total settlement should be computed at a sufficient number of points to establish the overall settlement pattern. From this pattern, the differential settlements can be determined and compared with recommended allowable values.

5-9. MASS CONCRETE STRUCTURE.

Mass concrete structures are uniquely designed and constructed to meet the needs of a particular project. These structures vary in size, shape,

TABLE 5-1. Summary of Finite Element Programs

Program (1)	Capabilities							
	2D and 3D solid elements (2)	Boundary elements (3)	Carck elements (4)	Linear elastic anisotropic (5)	Nonlinear elastic (6)	Plasticity (7)	Viscoelastic or creep (8)	Interactive graphics (9)
ABAQUS	X			X	X	X	X	X
ANSYS	X		X	X	X	X	X	X
APPLE-SAP	X	X		X				
ASKA	X		X	X	X	X	X	X
BEASY	X	X				X		
BERSAFE		X	X	X	X	X	X	X
BMINES	X		X	X	X	X	X	X
DIAL	X	X	X	X	X	X	X	X
MCAUTO STRUDL	X			X				X
MSC/NASTRAN	X		X	X	X	X	X	X
PAFEC	X	X	X	X	X	X	X	X
SAP(WES)	X			X				X
E3SAP	X			X				X
NONSAP	X		X	X	X	X		X
TITUS	X	X	X	X	X	X		X

and intended function between projects. As a result, the magnitude of differential settlement that can be tolerated must be established for each structure. Specifications for the allowable magnitudes of differential deformation/settlement that can be tolerated require the collective efforts of structural and geotechnical professionals, working together as a team. The magnitude of allowable differential movement should be sufficiently low so as to prevent the development of shear and/or tensile stresses within the structure in excess of tolerable limits and to ensure the proper functioning of movable features such as lock and flood control gates. For mass concrete structures founded on soft rock, where the modulus of deformation of the rock is significantly less than the elastic modulus of the concrete, there is a tendency for the foundation rock to expand laterally, thus producing additional tensile stresses along the base of the foundation. Deere, et al. (1967) suggested the following criteria for evaluating the significance of the ratios between the modulus of deformation of the rock (E_{dr}) and the elastic modulus of the concrete (E_c).

A. If $E_{dr}/E_c > 0.25$, then the foundation rock modulus has little effect on stresses generated within the concrete mass.

B. If $0.06 < E_{dr}/E_c < 0.25$, the foundation rock modulus becomes more significant with respect to stresses generated in the concrete structure. The significance increases with decreasing modulus ratio values.

C. If $E_{dr}/E_c < 0.06$, then the foundation rock modulus almost completely dominates the stresses generated within the concrete. Allowable magnitudes of deformation, in terms of settlement heave, lateral movement, or angular distortion for hydraulic structures, should be established by the design team and follow CECW-ED guidance.

SECTION IV. TREATMENT METHODS

5-10. GENERAL.

In design cases where the magnitudes of differential deformation/settlement exceed allowable values, the team of structural and geotechnical professionals charged with the responsibility of foundation design must make provisions for either reducing the magnitude of differential movement or design the structure to accommodate the differential deformation. A discussion of the latter option is beyond the scope of this manual. There are two approaches available for reducing the magnitude of differential

deformation/settlement: improve the rock mass deformation characteristics and/or modification of the foundation design.

5-11. ROCK MASS IMPROVEMENT.

Rock mass improvement techniques refer to techniques that enhance the ability of a rock mass to resist deformation when subjected to an increase in stress. The two techniques that are available include rock reinforcement and consolidation grouting. As a rule, techniques for increasing the modulus of deformation of a rock mass are limited to special cases where only relatively small reductions in deformation are necessary to meet allowable deformation/settlement requirements.

A. Rock Reinforcement. Rock reinforcement (i.e., rock bolts, rock anchor, rock tendon, etc.) is primarily used to enhance the stability of structures founded on rock. However, in specialized cases, constraint offered by a systematic pattern of rock reinforcement can be effective in reducing structural movement or translations (e.g., rotational deformations of retaining structures). Guidance for rock reinforcement systems is provided in Chapter 9.

B. Consolidation Grouting. Consolidation grouting refers to the injection of cementitious grouts into a rock mass for the primary purpose of increasing the modulus of deformation and/or shear strength. The enhancement capabilities of consolidation grouting depend upon rock mass conditions. Consolidation grouting to increase the modulus of deformation is more beneficial in highly fractured rock masses with a predominant number of open joints. Before initiating a consolidation grouting program, a pilot field study should be performed to evaluate the potential enhancement. The pilot field study should consist of trial grouting a volume of rock mass representative of the rock mass to be enhanced. In-situ deformation tests (discussed in Chapter 4) should be performed before and after grouting to evaluate the degree of enhancement achieved. Guidance pertaining to consolidation grouting is provided in EM 1110-2-3506 and Technical Report REMR-GT-8 (Dickinson 1988).

5-12. FOUNDATION DESIGN MODIFICATIONS.

The most effective means of reducing differential deformation/settlement are through modification of the foundation design. A variety of viable modifications is possible, but all incorporate one or more of three basic concepts: reduce stresses applied to the foundation rock; redistribute the applied stresses to stiffer and more competent rock strata; and, in cases involving flexible foundations, reduce maximum deformations by increasing the foundation stiffness. The choice of concept incorporated into the final design depends on the foundation rock conditions, structural considerations, and associated cost, and should be accomplished by the design team in accordance with CECW-ED guidance.

CHAPTER 6

BEARING CAPACITY

6-1. SCOPE.

This chapter provides guidance for the determination of the ultimate and allowable bearing stress values for foundations on rock. The chapter is subdivided into four sections with the following general topic areas: modes and examples of bearing capacity failures; methods for computing bearing capacity; allowable bearing capacity; and treatment methods for improving bearing capacity.

6-2. APPLICABILITY.

A. Modes of failure, methods for estimating the ultimate and allowable bearing capacity, and treatments for improving bearing capacity are applicable to structures founded directly on rock or shallow foundations on rock with depths of embedments less than four times the foundation width. Deep foundations such as piles, piers, and caissons are not addressed.

B. As a rule, the final foundation design is controlled by considerations such as deformation/settlement, sliding stability, or overturning rather than by bearing capacity. Nevertheless, the exceptions to the rule, as well as prudent design, require that the bearing capacity be evaluated.

SECTION I. FAILURE MODES

6-3. GENERAL.

Bearing capacity failures of structures founded on rock masses are dependent upon joint spacing with respect to foundation width, joint orientation, joint condition (open or closed), and rock type. Figure 6-1 illustrates typical failure modes according to rock mass conditions as modified from suggested modes by Sowers (1979) and Kulhawy and Goodman (1980). Prototype failure modes may actually consist of a combination of modes. For convenience of discussion, failure modes will be described according to four general rock mass conditions: intact, jointed, layered, and fractured.

6-4. INTACT ROCK MASS.

For the purpose of bearing capacity failures, intact rock refers to a rock mass with typical discontinuity spacing (Sterm in Figure 6-1) greater than four to five times the width (B term in Figure 6-1) of the foundation. As a rule, joints are so widely spaced that joint orientation and condition are of little importance. Two types of failure modes are possible depending on rock type. The two modes are local shear failure and general wedge failure associated with brittle and ductile rock, respectively.

A. Brittle Rock. A typical local shear failure is initiated at the edge of the foundation as localized crushing (particularly at edges of rigid foundations) and develops into patterns of wedges and slip surfaces. The slip surfaces do not reach the ground surface, however, ending somewhere in the rock mass. Localized shear failures are generally associated with brittle rock that exhibit significant post-peak strength loss (Figure 6-1A).

B. Ductile Rock. General shear failures are also initiated at the foundation edge, but the slip surfaces develop into well defined wedges which extend to the ground surface. General shear failures are typically associated with ductile rocks which demonstrate post-peak strength yield (Figure 6-1B).

6-5. JOINTED ROCK MASS.

Bearing capacity failures in jointed rock masses are dependent on discontinuity spacing, orientation, and condition.

A. Steeply Dipping and Closely Spaced Joints. Two types of bearing capacity failure modes are possible for structures founded on rock masses in which the predominant discontinuities are steeply dipping and closely spaced, as illustrated in Figure 6-1C and 6-1D. Discontinuities that are open (Figure 6-1C) offer little lateral restraint. Hence, failure is initiated by the compressive failure of individual rock columns. Tightly closed discontinuities (Figure 6-1D) on the other hand, provide lateral restraint. In such cases, general shear is the likely mode of failure.

B. Steeply Dipping and Widely Spaced Joints. Bearing capacity failures for rock

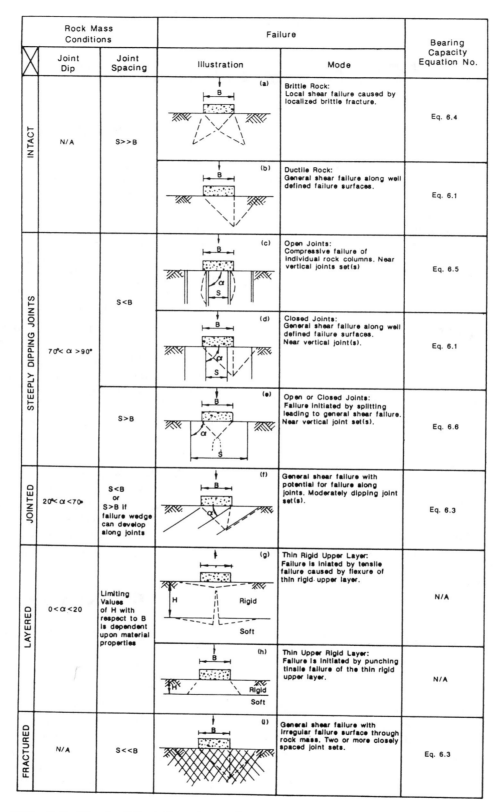

FIG. 6-1. Typical Bearing Capacity Failure Modes Associated with Various Rock Mass Conditions

masses with steeply dipping joints and with joint spacing greater than the width of the foundation (Figure 6-1E) are likely to be initiated by splitting that eventually progresses to the general shear mode.

C. Dipping Joints. The failure mode for a rock mass with joints dipping between 20 to 70 deg. with respect to the foundation plane is likely to be general shear (Figure 6-1F). Furthermore, since the discontinuity represents major planes of weakness, a favorably oriented discontinuity is likely to define at least one surface of the potential shear wedge.

6-6. LAYERED ROCK MASS.

Failure modes of multilayered rock masses, with each layer characterized by different material properties, are complicated. Failure modes for two special cases, however, have been identified (Sowers 1979). In both cases the founding layer consists of a rigid rock underlain by a soft highly deformable layer, with bedding planes dipping at less than 20 deg. with respect to the foundation plane. In the first case (Figure 6-1G), a thick rigid layer overlies the soft layer; while in the second case (Figure 6-1H) the rigid layer is thin. In both cases, failure is initiated by tensile failure. However, in the first case, tensile failure is caused by flexure of the rigid thick layer; while in the second case, tensile failure is caused by punching through the thin rigid upper layer. The limiting thickness of the rigid layer in both cases is controlled by the material properties of each layer.

6-7. HIGHLY FRACTURED ROCK MASSES.

A highly fractured rock mass is one that contains two or more discontinuity sets with typical joint spacings that are small with respect to the foundation width (Figure 6-1I). Highly fractured rock behaves in a manner similar to dense cohesionless sands and gravels. As such, the mode of failure is likely to be general shear.

6-8. SECONDARY CAUSES OF FAILURE.

In addition to the failure of the foundation rock, aggressive reactions within the rock mineralogy or with ground water or surface water chemistry can lead to bearing capacity failure. Examples include: loss of strength with time typical of some clay shales; reduction of load bearing cross-section caused by chemical reaction between the foundation element and the ground water or surface water; solution-susceptible rock materials; and additional stresses imposed by swelling minerals. Potential secondary causes should be identified during the site investigation phase of the project. Once the potential causes have been identified and addressed, their effects can be minimized.

SECTION II. METHODS FOR COMPUTING BEARING CAPACITY

6-9. GENERAL.

There are a number of techniques available for estimating the bearing capacity of rock foundations. These techniques include analytical methods, traditional bearing capacity equations, and field load tests. Of the various methods, field load tests are the least commonly used for two reasons. First, as discussed in Chapter 4, field load tests, such as the plate bearing test, are expensive. Second, although the test provides information as to the load that will cause failure, there still remains the question of scale effects.

6-10. DEFINITIONS.

Two terms used in the following discussions require definition. They are the ultimate bearing capacity and allowable bearing value. Definitions of the terms are according to the American Society for Testing and Materials.

A. Ultimate Bearing Capacity. The ultimate bearing capacity is defined as the average load per unit area required to produce failure by rupture of a supporting soil or rock mass.

B. Allowable Bearing Capacity Value. The allowable bearing capacity value is defined as the maximum pressure that can be permitted on a foundation soil (rock mass), giving consideration to all pertinent factors, with adequate safety against rupture of the soil mass (rock mass) or movement of the foundation of such magnitude that the structure is impaired. Allowable bearing values will be discussed in Section III of this chapter.

6-11. ANALYTICAL METHODS.

The ultimate bearing capacity may be implicitly estimated from a number of analytical methods. The more convenient of these methods include the finite element and limit equilibrium methods.

A. Finite Element Method. The finite element method is particularly suited to analyze foundations with unusual shapes and/or unusual loading conditions as well as in situations where the foundation rock is highly variable. For example, the potential failure modes for the layered foundation rock cases illustrated in Figures 6-1g and 6-1h will require consideration of the interactions between the soft and rigid rock layers as well as between the rigid rock layer and the foundation. The primary disadvantage of the finite element method is that the method does not provide a direct solution for the ultimate bearing capacity. Such solutions require an analyses of the resulting stress distributions with respect to a suitable failure criterion. In addition to the method's ability to address complex conditions, the primary advantage is that the method provides direct solutions for deformation/settlement.

B. Limit Equilibrium. The limit equilibrium method is applicable to bearing capacity failures defined by general wedge type shear, such as illustrated in Figures 6-1b, 6-1d, 6-1f, and 6-1i. The limit equilibrium method, as applied to sliding stability, is discussed in Chapter 7. Although the principals are the same as in sliding stability solutions, the general form of the equations presented in Chapter 7 needs to be cast in a form compatible with bearing capacity problems. The ultimate bearing capacity corresponds to the foundation loading condition necessary to cause an impending state of failure (i.e., the loading case where the factor of safety is unity).

6-12. BEARING CAPACITY EQUATIONS.

A number of bearing capacity equations are reported in the literature that provide explicit solutions for the ultimate bearing capacity. As a rule, the equations represent either empirical or semi-empirical approximations of the ultimate bearing capacity and are dependent on the mode of potential failure as well as, to some extent, material properties. In this respect, selection of an appropriate equation must anticipate likely modes of potential failure. The equations recommended in the following discussions are presented according to potential modes of failure. The appropriate equation number for each mode of failure is given in Figure 6-1.

A. General Shear Failure. The ultimate bearing capacity for the general shear mode of failure can be estimated from the traditional Buisman-Terzaghi (Terzaghi 1943) bearing capacity expression as defined by Equation 6-1. Equation 6-1 is valid for long continuous foundations with length-to-width ratios in excess of 10.

$$q_{ult} = cN_c + 0.5\,\gamma B N_\gamma + \gamma D N_q \qquad (6\text{-}1)$$

where q_{ult} = the ultimate bearing capacity; γ = effective unit weight (i.e., submerged unit wt. if below water table) of the rock mass; B = width of foundation; D = depth of foundation below ground surface; and c = cohesion intercepts for the rock mass. The terms N_c, N_γ, and N_q are bearing capacity factors given by the following equations.

$$N_c = 2N_\phi^{1/2} (N_\phi + 1) \qquad (6\text{-}2a)$$

$$N_\gamma = N_\phi^{1/2} (N_\phi^2 - 1) \qquad (6\text{-}2b)$$

$$N_q = N_\phi^2 \qquad (6\text{-}2c)$$

$$N_\phi = \tan^2 (45 + \phi/2) \qquad (6\text{-}2d)$$

where ϕ = angle of internal friction for the rock mass. Equation 6-1 is applicable to failure modes in which both cohesion and frictional shear strength parameters are developed. As such, Equation 6-1 is applicable to failure modes illustrated in Figures 6-1b and 6-1d.

B. General Shear Failure Without Cohesion. In cases where the shear failure is likely to develop along planes of discontinuity or through highly fractured rock masses, such as illustrated in Figures 6-1f and 6-1i, cohesion cannot be relied upon to provide resistance to failure. In such cases the ultimate bearing capacity can be estimated from the following equation:

$$q_{ult} = 0.5\,\gamma B N_\gamma + \gamma D N_q \qquad (6\text{-}3)$$

All terms are as previously defined.

C. Local Shear Failure. Local shear failure represents a special case where failure surfaces start to develop but do not propagate to the surface as illustrated in Figure 6-1a. In this respect, the depth of embedment contributes little to the total bearing capacity stability. An expression for the ultimate bearing capacity applicable to localized shear failure can be written as:

$$q_{ult} = cN_c + 0.5\gamma B N_\gamma \qquad (6\text{-}4)$$

All terms are as previously defined.

D. Correction Factors. Equations 6-1, 6-3, and 6-4 are applicable to long continuous foundations with length-to-width ratios (L/B) greater than 10. Table 6-1 provides correction factors for circular

TABLE 6-1. Correction Factors (after Sowers 1979)

Foundation shape (1)	C_c N_c correction (2)	$C\gamma$ $N\gamma$ correction (3)
Circular	1.2	0.70
Square	1.25	0.85
Rectangular		
L/B = 2	1.12	0.90
L/B = 5	1.05	0.95
L/B = 10	1.00	1.00

and square foundations, as well as rectangular foundations with L/B ratios less than ten. The ultimate bearing capacity is estimated from the appropriate equation by multiplying the correction factor by the value of the corresponding bearing capacity factor. Correction factors for rectangular foundations with L/B ratios other than 2 or 5 can be estimated by linear interpolation.

E. Compressive Failure. Figure 6-1c illustrates a case characterized by poorly constrained columns of intact rock. The failure mode in this case is similar to unconfined compression failure. The ultimate bearing capacity may be estimated from Equation 6-5.

$$q_{qult} = 2\ c \tan (45 + \phi/2) \qquad (6\text{-}5)$$

All parameters are as previously defined.

F. Splitting Failure. For widely spaced and vertically oriented discontinuities, failure generally initiates by splitting beneath the foundation as illustrated in Figure 6-1e. In such cases, Bishnoi (1968) suggested the following solutions for the ultimate bearing capacity:

For circular foundations:

$$q_{ult} = JcN_{cr} \qquad (6\text{-}6a)$$

For square foundations:

$$q = 0.85JcN_{cr} \qquad (6\text{-}6b)$$

For continuous strip foundations for $L/B \le 32$:

$$q_{ult} = JcN_{cr}/(2.2 + 0.18\ L/B) \qquad (6\text{-}6c)$$

where J = correction factor dependent upon thickness of the foundation rock and width of foundation;

and L = length of the foundation. The bearing capacity factor N_{cr} is given by:

$$N_{cr} = \frac{2N_\phi^2}{1 + N_\phi}\ (\cot\phi)\ (S/B)\ \left(1 - \frac{1}{N_\phi}\right) \qquad (6\text{-}6d)$$

$$- N\phi\ (\cot\phi) + 2N\phi^{1/2}$$

All other terms are as previously defined. Graphical solutions for the correction factor (J) and the bearing capacity factor (N_{cr}) are provided in Figures 6-2 and 6-3, respectively.

G. Input Parameters. The bearing capacity equations discussed above were developed from considerations of the Mohr-Columb failure criteria. In this respect, material property input parameters are limited to two parameters: the cohesion intercept (c) and the angle of internal friction (ϕ). Guidance for selecting design shear strength parameters is provided in Chapter 4. However, since rock masses generally provide generous margins of safety against bearing capacity failure, it is recommended that initial values of c and ϕ selected for assessing bearing capacity be based on lower bound estimates. While inexpensive techniques are available on which to base lower bound estimates of the friction angle, no inexpensive techniques are available for estimating lower bound cohesion values applicable to rock masses. Therefore, for computing the ultimate bearing capacity of a rock mass,

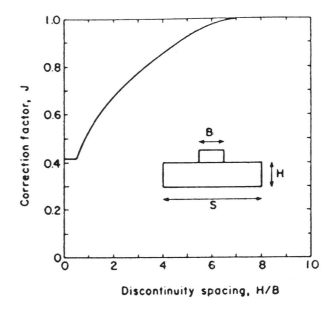

FIG. 6-2. Correction Factor for Discontinuity Spacing with Depth (after Bishnoi 1968)

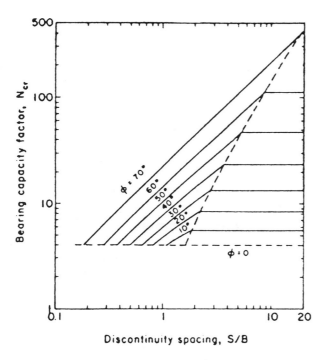

FIG. 6-3. Bearing Capacity Factor for Discontinuity Spacing (after Bishnoi 1968)

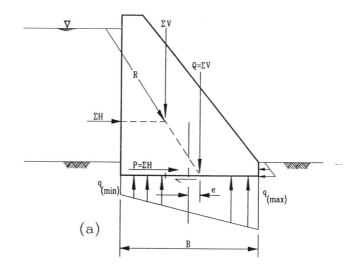

(a)

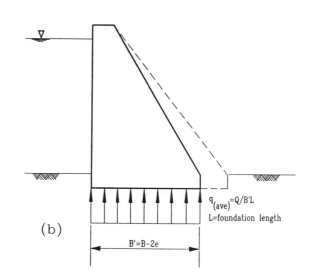

(b)

FIG. 6-4. Typical Eccentrically Loaded Structure Foundation

the lower bound value of cohesion may be estimated from the following equation.

$$c = \frac{q_u \, (s)}{2 \tan\left(45 + \dfrac{\phi}{2}\right)} \quad (6\text{-}7a)$$

where q_u = unconfined compressive strength of the intact rock from laboratory tests.

$$s = \exp\frac{(RMR - 100)}{9} \quad (6\text{-}7b)$$

All other parameters are as previously defined.

6-13. ECCENTRIC LOAD ON A HORIZONTAL FOUNDATION.

Eccentric loads acting on foundations effectively reduce the bearing capacity. Figure 6-4a illustrates a typical structure subjected to an eccentric load. In order to prevent loss of rock/structure contact at the minimum stress edge of the foundation (Figure 6-4a), the structure must be designed so that the resultant of all forces acting on the foundations passes through the center one-third of the foundation. As indicated in Figure 6-4a, the stress distribution can be approximated by linear relationship.

Equations 6-8a and 6-8b define the approximate maximum and minimum stress, respectively.

$$q_{(max)} = \frac{Q}{B}\left(1 + \frac{6e}{B}\right) \quad (6\text{-}8a)$$

$$q_{(min)} = \frac{Q}{B}\left(1 - \frac{6e}{B}\right) \quad (6\text{-}8b)$$

where $q_{(max)}$ = maximum stress; $q_{(min)}$ = minimum stress; Q = vertical force component of the resultant of all forces acting on the structure; B = foundation width; and e = distance from the center of the foundation to the vertical force component Q. The ultimate bearing capacity of the foundation can be approximated by assuming that the vertical force component Q is uniformly distributed across a reduced effective foundation width as indicated in Figure 6-4b. The effective width is defined by the following equation.

$$B' = B - 2e \qquad (6-9)$$

The effective width (B') is used in the appropriate bearing capacity equation to calculate the ultimate bearing capacity.

6-14. SPECIAL DESIGN CASES.

The bearing capacity equations discussed above are applicable to uniformly loaded foundations situated on planar surfaces. Frequently, designs suited to the particular requirements of a project require special considerations. Special design cases for which solutions of the ultimate bearing capacity are readily available are summarized in Figure 6-5. As indicated in Figure 6-5, these special cases include: inclined loads, inclined foundations, and foundations along or near slopes. Guidance for these special cases is provided in EM 1110-2-2502 and the NAVDOCKS DM-7. Ultimate bearing capacity solutions for special design cases should be in keeping with the modes of failure summarized in Figure 6-1.

SECTION III. ALLOWABLE BEARING CAPACITY VALUE

6-15. GENERAL.

The allowable bearing capacity value is defined in paragraph 6-10B. In essence, the allowable bearing capacity is the maximum limit of bearing stress that is allowed to be applied to the foundation rock. This limiting value is intended to provide a sufficient margin of safety with respect to bearing failures and deformation/settlement. Nevertheless, a prudent design dictates that, once the allowable bearing capacity value has been determined, a separate calculation be performed to verify that the allowable differential deformation/settlement is not exceeded.

6-16. DETERMINATION.

There are at least three approaches for determining allowable bearing capacity values. First, the allowable value may be determined by applying a

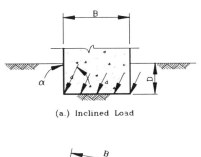

(a.) Inclined Load

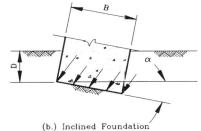

(b.) Inclined Foundation

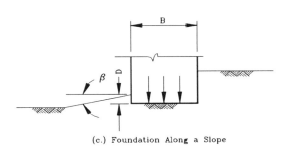

(c.) Foundation Along a Slope

FIG. 6-5. Special Foundation Design Cases

suitable factor of safety to the calculated ultimate bearing capacity. The selection of final allowable bearing values used in design of hydraulic structures must be based on the factor of safety approach in which all site-specific conditions and unique problems of such structures are considered. Second, allowable values may be obtained from various building codes. However, building codes, in general, apply only to residential or commercial buildings and are not applicable to the unique problems of hydraulic structures. Finally, allowable values may be obtained from empirical correlations. As a rule, empirical correlations are not site-specific and hence should be used only for preliminary design and/or site evaluation purposes. Regardless of the approach used, the allowable value selected for final design must not exceed the value obtained from the factor of safety considerations discussed in paragraph 6-16A.

A. Factor of Safety. The allowable bearing capacity value, q_a, based on the strength of the rock mass is defined as the ultimate bearing capacity, q_{ult}, divided by a factor of safety (FS):

$$q_a = q_{ult}/FS \qquad (6\text{-}10)$$

The average stress acting on the foundation material must be equal to or less than the allowable bearing capacity according to the following equation.

$$Q/BL \le q_a \qquad (6\text{-}11)$$

For eccentrically loaded foundations the B' value (i.e. Equation 6-9) is substituted for the B term in Equation 6-11. The factor of safety considers the variability of the structural loads applied to the rock mass, the reliability with which foundation conditions have been determined, and the variability of the potential failure mode. For bearing capacity problems of a rock mass, the latter two considerations are the controlling factors. For most structural foundations, the minimum acceptable factor of safety is 3, with a structural load comprised of the full dead load plus the full live load.

B. Building Codes. Allowable bearing capacity values that consider both strength and deformation/settlement are prescribed in local and national building codes. Local codes are likely to include experience and geology within their jurisdiction while national codes are more generic. For example, a local code will likely specify a particular rock formation such as "well-cemented Dakota sandstone," while a national code may use general terminology such as "sedimentary rock in sound condition." As a rule, allowable values recommended by the building codes are conservative.

C. Empirical Correlations. Peck, Hanson, and Thornburn (1974) suggested an empirical correlation between the allowable bearing capacity stress and the RQD, as shown in Figure 6-6. The correlation is intended for a rock mass with discontinuities that "are tight or are not open wider than a fraction of an inch."

6-17. STRUCTURAL LIMITATIONS.

The maximum load that can be applied to a rock foundation is limited by either the rock's ability to sustain the force without failure or excessive settlement, or the ability of the substructure to sustain the load without failure or excessive deformation. In some cases the structural design of the foundation element will dictate the minimum element size and, consequently, the maximum contact stress on the rock. For typical concrete strengths in use today, the strength of the concrete member is significantly less than the bearing capacity of many rock masses.

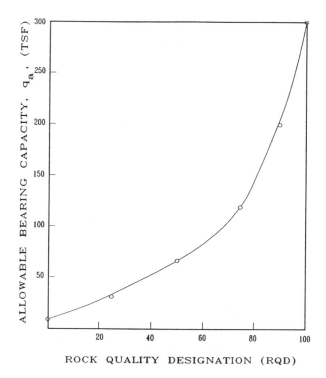

FIG. 6-6. Allowable Contact Pressure on Jointed Rock

SECTION IV. TREATMENT METHODS

6-18. GENERAL.

Treatment methods for satisfying bearing capacity requirements are essentially the same as those for satisfying deformation/settlement requirements discussed in Chapter 5. In addition to the previously discussed methods, an examination of the general ultimate bearing capacity equation (i.e., Equation 6-1) indicates the importance of two parameters not directly related to deformability. These two parameters are the effective unit weight of the foundation rock and the depth of the foundation below the ground surface.

6-19. EFFECTIVE UNIT WEIGHT.

For foundations below the water table, the effective unit weight is the unit weight of the foundation rock minus the unit weight of water (i.e., submerged unit weight of the rock). Hence, foundations located above the water table will develop significantly more resistance to potential bearing capacity failures than foundations below the water table.

6-20. FOUNDATION DEPTH.

Foundations constructed at greater depths may increase the ultimate bearing capacity of the foundation. The improved capacity is due to a greater passive resisting force and a general increase in rock mass strength with depth. The increased lithostatic pressure closes discontinuities, and the rock mass is less susceptible to surficial weathering. Occasionally, deeper burial may not be advantageous. A region with layers of differing rock types may contain weaker rock at depth. In such an instance, a strong rock might overlie a layer such as mudstone, or, if in a volcanic geology, it might be underlain by a tuff or ash layer. In these instances, deeper burial may even decrease the bearing capacity. The geologic investigation will determine this possibility.

CHAPTER 7

SLIDING STABILITY

7-1. SCOPE.

This chapter provides guidance for assessing the sliding stability of laterally loaded structures founded on rock masses. Examples of applicable structures include gravity dams, coffer dams, flood walls, lock walls, and retaining structures. The chapter is divided into three sections to include: modes of failure, methods of analysis, and treatment methods.

SECTION I. MODES OF FAILURE

7-2. GENERAL.

Paths along which sliding can occur will be confined to the foundation strata; pass through both the foundation strata and the structure; or just pass through the structure. This chapter addresses sliding where the failure path is confined to the foundation strata or at the interface between the strata and the structure's foundation. Although complex, foundation-structure sliding failure or sliding failure through the structure are conceptually possible and must be checked, such failures are likely to occur only in earth structures (e.g., embankments). The analyses of these later two failure modes are addressed in EM 1110-2-1902.

7-3. POTENTIAL FAILURE PATHS.

Potential failure paths along which sliding may occur can be divided into five general categories as illustrated in Figure 7-1.

A. Failure Along Discontinuities. Figure 7-1A illustrates a mode of potential failure where the failure path occurs along an unfavorably oriented discontinuity. The mode of failure is kinematically possible in cases where one or more predominate joint sets strike roughly parallel to the structure and dip in the upstream direction. The case is particularly hazardous with the presence of an additional joint set striking parallel to the structure and dipping downstream. In the absence of the additional joint set, failure is generally initiated by a tensile failure at the heel of the structure. Where

possible, the structure should be aligned in a manner that will minimize the development of this potential mode of failure.

B. Combined Failure. A combined mode of failure is characterized by situations where the failure path can occur both along discontinuities and through intact rock as illustrated in Figure 7-1B. Conceptually, there are any number of possible joint orientations that might result in a combined mode of failure. However, the mode of failure is more likely to occur in geology where the rock is horizontally or near horizontally bedded and the intact rock is weak.

C. Failure Along Interface. In cases where structures are founded on rock masses containing widely spaced discontinuities, none of which are unfavorably oriented, the potential failure path is likely to coincide with the interface between the structure and the foundation strata. The interface mode of failure is illustrated in Figure 7-1C.

D. Generalized Rock Mass Failure. In the generalized rock mass mode of failure, the failure path is a localized zone of fractured and crushed rock rather than well defined surfaces of discontinuity. As implied in Figure 7-1D, a generalized rock mass failure is more likely to occur in highly fractured rock masses.

E. Buckling Failure. Figure 7-1E illustrates a conceptual case where failure is initiated by buckling of the upper layer of rock downstream of the structure. Rock masses conducive to buckling type failures would contain thin, horizontally bedded rock in which the parent rock is strong and brittle. Although no case histories have been recorded where buckling contributed to or caused failure, the potential for a buckling failure should be addressed where warranted by site conditions.

SECTION II. METHODS OF ANALYSIS

7-4. GENERAL APPROACH.

The guidance in this chapter is based on conventional geotechnical principles of limit equilibrium. The basic principle of this method applies the factor of safety to the least known conditions affect-

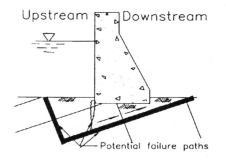

(a.) Failure Along Discontinuities

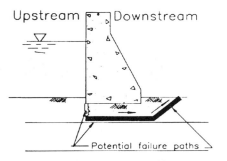

(b.) Combined Failure

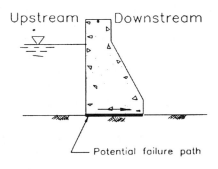

(c.) Failure Along Interface

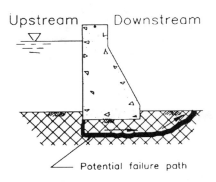

(d.) Generalized Rock Mass Failure

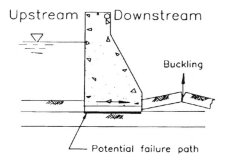

(e.) Failure Initiated By Buckling

FIG. 7-1. Potential Failure Paths

ing sliding stability, that is, the material shear strength. Mathematically, the basic principle is expressed as:

$$\tau = \frac{\tau_f}{FS} \qquad (7\text{-}1)$$

in which τ is the limiting (applied) shear stress required for equilibrium and τ_f is the maximum available shear strength that can be developed. The ratio of these two quantities, expressed by Equation 7-2, is called the factor of safety.

$$FS = \tau_f / \tau \qquad (7\text{-}2)$$

The maximum available shear strength τ_f is defined by the Mohr-Coulomb failure criterion. Procedures for selecting the appropriate shear strength parameters c and ϕ are discussed in Chapter 4.

7-5. CONDITIONS FOR STABILITY.

According to this method, the foundation is stable with respect to sliding when, for any potential slip surface, the resultant of the applied shear stresses required for equilibrium is smaller than the maximum shear strength that can be developed. A factor of safety approaching unity for any given potential slip surface implies failure by sliding is impending. The surface along which sliding has the greatest probability of occurring is the surface that results in the smallest factor of safety. This surface is referred to as the potential critical failure surface.

7-6. ASSUMPTIONS.

As in any mathematical expression that attempts to model a geologic phenomenon, the limit equilibrium method requires the imposition of certain simplifying assumptions. Assumptions invariably translate into limitations in application. Limit equilibrium methods will provide an adequate assessment of sliding stability provided that sound engineering judgment is exercised. This judgment requires a fundamental appreciation of the assumptions involved and the resulting limitations imposed. The following discussion emphasizes the more important assumptions and limitations.

A. Failure Criterion. Conventional limit equilibrium solutions for assessing sliding stability incorporate the linear Mohr-Coulomb failure criterion (see Figure 4-5) for estimating the maximum available shear strength (τ_f). It is generally recognized that failure envelopes for all modes of rock failure are, as a rule, non-linear. As discussed in Chapter 4, imposition of a linear criterion for failure, as applied to rock, requires experience and judgment in selecting appropriate shear strength parameters.

B. Two-dimensional Analysis. The method presented in this chapter is two-dimensional in nature. In most cases, problems associated with sliding in rock masses involve the slippage of three-dimensional wedges isolated by two or more discontinuities and the ground surface. In such cases, a two-dimensional analysis generally results in a conservative assessment of sliding stability. It is possible for a two-dimensional analysis to predict an impending failure, where in reality the assumed failure mechanism is kinematically impossible.

C. Failure Surface. The stability equations are based on an assumed failure surface consisting of one or more planes. Multiplane surfaces form a series of wedges which are assumed to be rigid. The analysis follows the method of slices approach common to limit equilibrium generalized slip surfaces used in slope stability analysis (e.g., see Janbu 1973). Slices are taken at the intersection of potential failure surface planes. Two restrictions are imposed by the failure surface assumptions. First, the potential failure surface underlying the foundation element is restricted to one plane. Second, planear

surfaces are not conducive to search routines to determine the critical potential failure surface. As a result, determination of the critical failure surface may require a large number of trial solutions, particularly in rock masses with multiple, closely spaced, joint sets.

D. Force Equilibrium. Equations for assessing stability were developed by resolving applied and available resisting stresses into forces. The following assumptions are made with respect to forces.

(1) Only force equilibrium is satisfied. Moment equilibrium is not considered. Stability with respect to overturning must be determined separately.

(2) In order to simplify the stability equations, forces acting vertically between wedges are assumed to be zero. Neglecting these forces generally results in a conservative assessment of sliding stability.

(3) Because only forces are considered, the effects of stress concentrations are unknown. Potential problems associated with stress concentrations must be addressed separately. The finite element method is ideally suited for this task.

E. Strain Compatibility. Considerations regarding displacements are excluded from the limit equilibrium approach. The relative magnitudes of the strain at failure for different foundation materials may influence the results of the sliding stability analysis. Such complex structure-foundation systems may require a more intensive sliding investigation than a limit equilibrium approach. In this respect, the effects of strain compatibility may require special interpretation of data from in-situ tests, laboratory tests, and finite element analyses.

F. Factor of Safety. Limit equilibrium solutions for sliding stability assume that the factor of safety of all wedges are equal.

7-7. ANALYTICAL TECHNIQUES FOR MULTI-WEDGE SYSTEMS.

A. General Wedge Equations. The general wedge equations are derived from force equilibrium of all wedges in a system of wedges defined by the geometry of the structure and potential failure surfaces. Consider the *i*th wedge in a system of wedges illustrated in Figure 7-2. The necessary geometry notation for the *i*th wedge and adjacent wedges are as shown (Figure 7-2). The origin of the coordinate system for the wedge considered is located in the lower left-hand corner of the

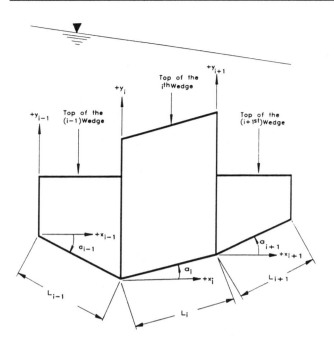

FIG. 7-2. Hypothetical *i*th Wedge and Adjacent Wedges Subject to Potential Sliding

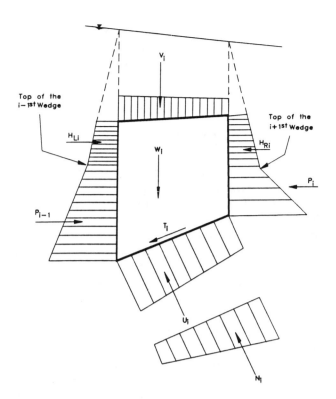

FIG. 7-3. Distribution of Pressures, Stresses and Resultant Forces Acting on Hypothetical *i*th Wedge

wedge. The x and y axes are horizontal and vertical respectively. Axes which are tangent (*t*) and normal (*n*) to the failure plane are oriented at an angle (α) with respect to the +x and +y axes. A positive value of α is a counterclockwise rotation, a negative value of α is a clockwise rotation. The distribution of pressures/stresses with resulting forces is illustrated in Figure 7-3. Figure 7-4 illustrates the free body diagram of the resulting forces. Summing the forces normal and tangent to the potential failure surface and solving for $(P_{i-1} - P_i)$ results in the following equation for the *i*th wedge:

$$(P_{i-1} - P_i) = \left[((W_i + V_i)\cos\alpha_i \right.$$

$$- U_i + (H_{Li} - H_{Ri})\sin\alpha_i$$

$$\frac{\tan\phi_i}{FS_i} - (H_{Li} - H_{Ri})\cos\alpha_i$$

$$\left. + (W_i + V_i)\sin\alpha_i + \frac{C_i}{FS_i} L_i \right]$$

$$\div \left[\cos\alpha_i - \sin\alpha_i \frac{\tan\phi_i}{FS_i} \right]$$

(7-3)

where

i = subscript notation for the wedge considered;

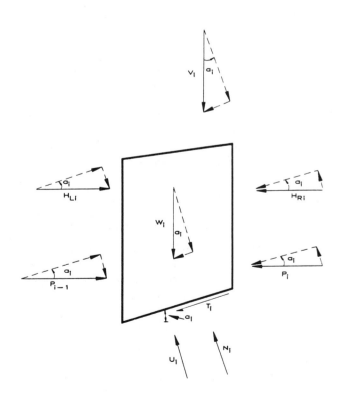

FIG. 7-4. Free Body Diagram of Hypothetical *i*th Wedge

P = horizontal residual forces acting between wedges as a result of potential sliding;

W = total weight of wedge to include rock, soil, concrete and water (do not use submerged weights);

V = any vertical force applied to the wedge;

α = angle of potential failure plane with respect to the horizontal ($-\alpha$ denotes downslope sliding, $+\alpha$ denotes upslope sliding);

U = uplift force exerted on the wedge at the potential failure surface;

H = in general, any horizontal force applied to the wedge (H_L and H_R refers to left and right hard forces as indicated in Figures 7-3 and 7-4);

L = length of the wedge along the potential failure surface;

FS = factor of safety;

c = cohesion shear strength parameter; and

ϕ = angle of internal friction.

B. Equilibrium Requirements. An inspection of Equation 7-3 reveals that for a given wedge there will be two unknowns (i.e., $(P_{i-1} - P_i)$ and FS). In a wedge system with n number of wedges, Equation 7-3 will provide n number of equations. Because FS is the same for all wedges, there will be $n + 1$ unknowns with n number of equations for solution. The solution for the factor of safety is made possible by a conditional equation establishing horizontal equilibrium of the wedge system. This equation states that the sum of the differences in horizontal residual forces $(P_{i-1} - P_i)$ acting between wedges must equal the differences in the horizontal boundary forces. Since boundary forces are usually equal to zero, the conditional equation is expressed as

$$\sum_{i=1}^{i=n} (P_{i-1} - P_i) = 0 \qquad (7\text{-}4)$$

where n = total number of wedges in the system.

C. Alternate Equation. An alternate equation for the implicit solution of the factor of safety for a system of n wedges is given below:

$$FS = \frac{\sum\limits_{i=1}^{i=n} \dfrac{C_i L_i \cos\alpha_i + (W_i + V_i - U_i \cos\alpha_i)\tan\phi_i}{n_{\alpha_i}}}{\sum\limits_{i=1}^{i=n} [H_i - (W_i + V_i)\tan\alpha_i]} \qquad (7\text{-}5a)$$

where

$$n_{\alpha_i} = \frac{1 - \dfrac{\tan\phi_i \tan\alpha_i}{FS}}{1 + \tan^2\alpha_i} \qquad (7\text{-}5b)$$

All other terms are as defined above. The derivation of Equations 7-5 follows that of Equations 7-3 and 7-4, except that forces are summed with respect to the x and y coordinates.

7-8. PRELIMINARY PROCEDURES.

Factor of safety solutions for a multi-wedge system containing a number of potential failure surfaces can result in a significant bookkeeping problem. For this reason, it is recommended that prior to the analytical solution for the factor of safety, the following preliminary procedures be implemented.

A. Define and identify, on a scale drawing, all potential failure surfaces based on the stratification, location, orientation, frequency, and distribution of discontinuities within the foundation material as well as the geometry, location, and orientation of the structure.

B. For each potential failure surface, divide the mass into a number of wedges. A wedge must be created each time there is a change in slip plane orientation and/or a change in shear strength properties. However, there can be only one structural wedge.

C. For each wedge draw a free body diagram that shows all the applied and resulting forces acting on that wedge. Include all necessary dimensions on the free body diagram. Label all forces and dimensions according to the appropriate parameter notations discussed above.

D. Prepare a table that lists all parameters to include shear strength parameters for each wedge in the system of wedges defining the potential slip mass.

7-9. ANALYTICAL PROCEDURES.

While both the general wedge equation and the alternate equation will result in the same calculated factor of safety for a given design case, the procedure for calculating that value is slightly different. Solutions for hypothetical example problems are provided in EM 1110-2-2200 and Nicholson (1983a).

A. General Wedge Method. The solution for the factor of safety using Equations 7-3 and

7-4 requires a trial-and-error procedure. A trial value for the factor of safety, FS, is inserted in Equation 7-3 for each wedge to obtain values of the differences in horizontal residual P forces acting between wedges. The differences in P forces for each wedge are then summed; a negative value indicates that the trial value of FS was too high and, conversely, a positive value indicates that the trial value of FS was too low. The process is repeated until the trial FS value results in an equality from Equation 7-4. The value of FS which results in an equality is the correct value for the factor of safety. The number of trial-and-error cycles can be reduced if trial values of FS are plotted with respect to the sum of the differences of the P forces. (See examples in EM 1110-2-2200 and Nicholson (1983a).)

B. Alternate Methods. Equations 7-5a and 7-5b, when expanded, can be used to solve for the factor of safety for a system containing one or more wedges. Since the n_α term, defined by Equation 7-5b, is a function of FS, the solution for FS requires an iterative process. An assumed initial value of FS is inserted into the $n\alpha$ term for each wedge in the expanded form of Equation 7-5a, and a new factor of safety is calculated. The calculated factor of safety is then inserted into the $n\alpha$ term. The process is repeated until the inserted value of FS equals the calculated value of FS. Convergence to within two decimal places usually occurs in three to four iteration cycles.

C. Comparison of Methods. The general wedge equation (Equation 7-3) was formulated in terms of the difference in horizontal boundary forces to allow the design engineer to solve directly for forces acting on the structure for various selected factors of safety. The procedure has an advantage for new structures in that it allows a rapid assessment of the horizontal forces necessary for equilibrium for prescribed factors of safety. The alternate equation (Equations 7-5a and 7-5b) solves directly for FS. Its advantage is in the assessment of stability for existing structures. Both equations are mathematically identical (Nicholson 1983a).

7-10. DESIGN CONSIDERATIONS.

Some special considerations for applying the general wedge equation to specific site conditions are discussed below.

A. Active Wedge. The interface between the group of active wedges and the structural wedge is assumed to be a vertical plane located at the heel of the structural wedge and extending to the base of the structural wedge. The magnitudes of the active forces depend on the actual values of the safety factor, the inclination angles (α) of the slip path, and the magnitude of the shear strength that can be developed. The inclination angles, corresponding to the maximum active residual P forces for each potential failure surface, can be determined by independently analyzing the group of active wedges for trial safety factors. In rock, the inclination may be predetermined by discontinuities in the foundation.

B. Structural Wedge. Discontinuities in the slip path beneath the structural wedge should be modeled by assuming an average slip-plane along the base of the structural wedge.

C. Passive Wedge. The interface between the group of passive wedges and the structural wedge is assumed to be a vertical plane located at the toe of the structural wedge and extending to the base of the structural wedge. The magnitudes of the passive residual P forces depend on the actual values of the safety factor, the inclination angles of the slip path, and the magnitude of shear strength that can be developed. The inclination angles, corresponding to the minimum passive residual P forces for each potential failure mechanism, can be estimated by independently analyzing the group of passive wedges for trial safety factors. When passive resistance is used, special considerations must be made. Removal of the passive wedge by future construction must be prevented. Rock that may be subjected to high-velocity water scouring should not be used unless amply protected. Also, the compressive strength of the rock layers must be sufficient to develop the wedge resistance. In some cases, wedge resistance should not be assumed without resorting to special treatment such as installing rock anchors.

D. Tension Cracks. Sliding analyses should consider the effects of cracks on the active side of the structural wedge in the foundation material due to differential settlement, shrinkage, or joints in a rock mass. The depth of cracking in cohesive foundation material can be estimated in accordance with the following equations.

$$d_c = \frac{2c_d}{\gamma} \tan\left(45 - \frac{\phi_d}{2}\right) \qquad (7\text{-}6a)$$

where

$$c_d = \frac{c}{FS} \qquad (7\text{-}6b)$$

$$\phi_d = \tan^{-1}\left(\frac{\tan\phi}{FS}\right) \qquad (7\text{-}6c)$$

The value (d_c) in a cohesive foundation cannot exceed the embedment of the structural wedge. The depth of cracking in massive, strong, rock foundations should be assumed to extend to the base of the structural wedge. Shearing resistance along the crack should be ignored and full hydrostatic pressure should be assumed to extend to the bottom of the crack. The hydraulic gradient across the base of the structural wedge should reflect the presence of a crack at the heel of the structural wedge.

E. Uplift Without Drains. The effects of seepage forces should be included in the sliding analysis. Analyses should be based on conservative estimates of uplift pressures. Estimates of uplift pressures on the wedges can be based on the following assumptions:

(1) The uplift pressure acts over the entire area of the base.

(2) If seepage from headwater to tailwater can occur across a structure, the pressure head at any point should reflect the head loss due to water flowing through a medium. The approximate pressure head at any point can be determined by the line-of-seepage method. This method assumes that the head loss is directly proportional to the length of the seepage path. The seepage path for the structural wedge extends from the upper surface (or internal ground-water level) of the uncracked material adjacent to the heel of the structure, along the embedded perimeter of the structural wedge, to the upper surface (or internal ground-water level) adjacent to the toe of the structure. Referring to Figure 7-5, the seepage distance is defined by points a, b, c, and d. The pressure head at any point is equal to the elevation head minus the product of the hydraulic gradient times the distance along the seepage path to the point in question. Estimates of pressure heads for the active and passive wedges should be consistent with those of the heel and toe of the structural wedge.

(3) For a more detailed discussion of the line-of-seepage method, refer to EM 1110-2-2502, Retaining and Flood Walls. For the majority of structural stability computations, the line-of-seepage is considered sufficiently accurate. However, there may be special situations where the flow net method is required to evaluate seepage problems.

F. Uplift With Drains. Uplift pressures on the base of the structural wedge can be reduced by foundation drains. The pressure heads beneath the structural wedge developed from the line-of-seepage analysis should be modified to reflect the effects of the foundation drains. The maximum pressure head along the line of foundation drains can be estimated from Equation 7-7:

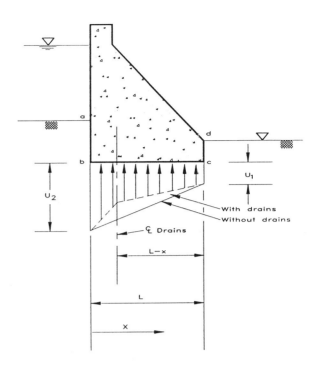

Pressure Head at Drains $= U_x = U_1 + R\left(\frac{L-x}{L}\right)(U_2 - U_1)$
U_1 = Pressure Head at Toe
U_2 = Pressure Head at Heel
R = Constant $\{100 - (25\% \rightarrow 50\%)\}$

FIG. 7-5. Uplift Pressures

$$U_x = U_1 + R\left(\frac{L - x}{L}\right)(U_2 - U_1) \qquad (7\text{-}7)$$

All parameters are defined in Figure 7-5. The uplift pressure across the base of the structural wedge usually varies from the undrained pressure head at the heel to the assumed reduced pressure head at the line of drains to the undrained pressure head at the toe, as shown in Figure 7-5. Uplift forces used for the sliding analyses should be selected in consideration of conditions that are presented in the applicable design memoranda. For a more detailed discussion of uplift under gravity dams, refer to EM 1110-2-2200, Gravity Dams.

G. Overturning. As stated previously, requirements for rotational equilibrium are not directly included in the general sliding stability equations. For some load cases, the vertical component of the resultant load will lie outside the kern of the base area, and a portion of the structural wedge will not be in contact with the foundation material. The sliding analysis should be modified for these load cases to reflect the following secondary effects due to coupling of sliding and overturning behavior.

(1) The uplift pressure on the portion of the base that is not in contact with the foundation material should be a uniform value which is equal to the maximum value of the hydraulic pressure across the base (except for instantaneous loads such as those due to seismic forces).

(2) The cohesive component of the sliding resistance should only include the portion of the base area that is in contact with the foundation material.

(3) The resultant of the lateral earth (soil) pressure is assumed to act at 0.38 of the wall height for horizontal or downward sloping backfills and at 0.45 of the wall height for upward sloping backfills.

(4) Cantilever or gravity walls on rock should be designed for at-rest earth pressures unless the foundation rock has an unusually low modulus.

7-11. SEISMIC SLIDING STABILITY.

The sliding stability of a structure for an earthquake-induced base motion should be checked by assuming the specified horizontal earthquake acceleration coefficient and the vertical earthquake acceleration coefficient, if included in the analysis, to act in the most unfavorable direction. The earthquake-induced forces on the structure and foundation wedges may then be determined by a quasi-static rigid body analysis. For the quasi-static rigid body analysis, the horizontal and vertical forces on the structure and foundation wedges may be determined by using the following equations:

$$H_{di} = M_i\ddot{X} + m_i\ddot{X} + H_i \qquad (7\text{-}8)$$

$$V_{di} = M_ig - m_i\ddot{y} \qquad (7\text{-}9)$$

where

H_d = horizontal forces acting on the structure and/or wedge;

V_d = vertical forces acting on the structure and or wedge;

M = mass of the structure and/or wedge (weight/g);

m = added mass of reservoir and/or adjacent soil/rock;

g = acceleration of gravity;

$\ddot{X}$ = horizontal earthquake acceleration coefficient;

and $\ddot{y}$ = vertical earthquake acceleration coefficient. The subscript i, H, and V terms are as defined previously.

A. Earthquake Acceleration. The horizontal earthquake acceleration coefficient can be obtained from seismic zone maps (ER 1110-2-1806) or, in the case where a design earthquake has been specified for the structure, an acceleration developed from analysis of the design earthquake. Guidance is being prepared for the latter type of analysis and will be issued in the near future; until then, the seismic coefficient method is the most expedient method to use. The vertical earthquake acceleration is normally neglected but can be taken as two-thirds of the horizontal acceleration if included in the analysis.

B. Added Mass. The added mass of the reservoir and soil can be approximated by Westergaard's parabola (EM 1110-2-2200) and the Mononobe-Okabe method (EM 1110-2-2502), respectively. The structure should be designed for a simultaneous increase in force on one side and decrease on the opposite side of the structure when such can occur.

C. Analytical Procedures. The analytical procedures for the seismic quasi-static analyses follows the procedures outlined in paragraphs 7-9A and 7-9B for the general wedge and alternate methods, respectively. However, the H_d and V_d terms are substituted for the H and W terms, respectively, in Equations 7-3 and 7-5a.

7-12. FACTOR OF SAFETY.

For major concrete structures (i.e., dams, lockwalls, basin walls that retain a dam embankment, etc.) the minimum required factor of safety for normal static loading conditions is 2.0. The minimum required factor of safety for seismic loading conditions is 1.3. Retaining walls on rock require a safety factor of 1.5; refer to EM 1110-2-2502 for a discussion of safety factors for floodwalls. Any relaxation of these values will be allowed only with the approval of CECW-E and should be justified by comprehensive foundation studies of such nature as to reduce uncertainties to a minimum.

SECTION III. TREATMENT METHODS

7-13. GENERAL.

Frequently, a sliding stability assessment of structures subjected to lateral loading results in an unacceptably low factor of safety. In such cases, a number of methods are available for increasing the resistance to sliding. An increase in sliding resistance may be achieved by one or a combination of three mechanistic provisions. The three provisions include: increasing the resisting shear strength by increasing the stress acting normal to the potential failure surface; increasing the passive wedge resistance; and providing lateral restraining forces.

7-14. INCREASE IN SHEAR STRENGTH.

The shear strength available to resist sliding is proportional to the magnitude of the applied stress acting normal to the potential slip surface. An increase in the normal stress may be achieved by either increasing the vertical load applied to the structural wedge and/or passive wedge(s) or by a reduction in uplift forces. The applied vertical load can be conveniently increased by increasing the mass of the structure or placing a berm on the downstream passive wedge(s). Installation of foundation drains and/or relief wells to relieve uplift forces is one of the most effective methods by which the stability of a gravity hydraulic structure can be increased.

7-15. INCREASE IN PASSIVE WEDGE RESISTANCE.

Resistance to sliding is directly influenced by the size of the passive wedge acting at the toe of the structure. The passive wedge may be increased by increasing the depth the structure is embedded in the foundation rock or by construction of a key. Embedment and keys are also effective in transferring the shear stress to deeper and frequently more competent rock.

7-16. LATERAL RESTRAINT.

Rock anchors inclined in the direction of the applied shear load provide a force component that acts against the applied shear load. Guidance for the design of anchor systems is discussed in Chapter 9 of this manual.

CHAPTER 8

CUT SLOPE STABILITY

8-1. SCOPE.

This chapter provides guidance for assessing the sliding stability of slopes formed by excavations in rock or of natural rock slopes altered by excavation activities. Typical examples of slopes cut in rock include: foundation excavations; construction of project access roads; and development of dam abutments, spillways, and tunnel portals. This chapter is divided into three sections according to the general topic areas of modes of failure, methods of assessing stability, and treatment methods and planning considerations.

SECTION I. MODES OF FAILURE

8-2. GENERAL.

The primary objectives of any rock excavation is to minimize the volume of rock excavated while providing an economical and safe excavation suitable for its intended function. The objectives of economy and safety, as a rule, involve the maximization of the angle of inclination of the slope while assuring stability. Stability assurance requires an appreciation for the potential modes of failure.

8-3. TYPES OF FAILURE MODES.

Because of its geometry, rock slopes expose two or more free surfaces. Thus, as a rule, constituent rock blocks contained within the rock mass have a relative high kinematic potential for instability. In this respect, the type of failure is primarily controlled by the orientation and spacing of discontinuities within the rock mass as well as the orientation of the excavation and the angle of inclination of the slope. The modes of failure that are controlled by the above factors can be divided into three general types: sliding, toppling, and localized sloughing. Each type of failure may be characterized by one or more failure mechanisms.

8-4. SLIDING FAILURE MODES.

Figure 8-1 illustrates seven failure mechanisms that may be associated with the sliding failure mode. While other failure mechanisms are concep-

tually possible, the seven mechanisms illustrated are representative of those mechanisms most likely to occur. The following discussions provide a brief description of the conditions necessary to initiate each of the sliding mechanisms.

A. Single Block/Single Sliding Plane. A single block with potential for sliding along a single plane (Figure 8-1A) represents the simplest sliding mechanism. The mechanism is kinematically possible in cases where at least one joint set strikes approximately parallel to the slope strike and dips toward the excavation slope. Failure is impending if the joint plane intersects the slope

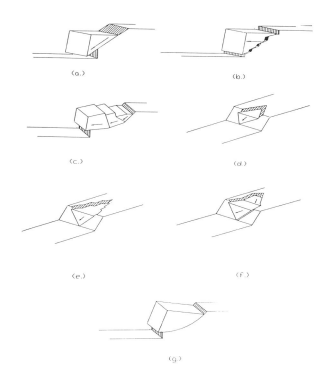

FIG. 8-1. Failure Mechanisms for Sliding Failure Mode: (a) Single Block with Single Plane; (b) Single Block with Stepped Planes; (c) Multiple Blocks with Multiple Planes; (d) Single Wedge with Two Intersecting Planes; (e) Single Wedge with Multiple Intersecting Planes; (f) Multiple Wedges with Multiple Intersecting Planes; and (g) Single Block with Circular Slip Path

plane and the joint dips at an angle greater than the angle of internal friction (ϕ) of the joint surface.

B. Single Block/Stepped Sliding Planes. Single block sliding along stepped planes (Figure 8-1B) is possible in cases where a series of closely spaced parallel joints strike approximately parallel to the excavation slope strike and dip toward the excavation slope. The parallel joints may or may not be continuous. However, at least one joint plane must intersect the slope plane. In the case of continuous parallel joints, a second set of joints is necessary. This second joint set must also strike more or less parallel to the slope, and the magnitude and direction of the joint dip angle must be such that the joint plane does not intersect the slope plane.

C. Multiple Blocks/Multiple Sliding Planes. Multiple blocks, sliding along multiple planes (Figure 8-1C) is the most complicated planar type of sliding. The mechanism is associated with two or more joint sets that strike approximately parallel to the slope strike and dip in the direction of the excavation slope. At least one of the joint planes must intersect the excavated slope plane. For a failure to occur, the dip angle of the joint defining the base of the uppermost block must be greater than the friction angle of the the joint surface. Furthermore, additional joints must be present that also strike approximately parallel to the strike of the excavated slope. These additional joints must either dip in a near vertical direction or dip steeply away from the slope plane.

D. Single Wedge/Two Intersecting Planes. Single wedge sliding (Figure 8-1D) can occur in rock masses with two or more sets of discontinuities whose lines of intersection are approximately perpendicular to the strike of the slope and dip toward the plane of the slope. In addition, this mode of failure requires that the dip angle of at least one joint-intersect is greater than the friction angle of the joint surfaces and that the line of joint intersection intersects the plane of the slope.

E. Single Wedge/Multiple Intersecting Planes. The conditions for sliding of a single wedge formed by the intersections of at least two discontinuity sets with closely spaced joints (Figure 8-1E) are essentially the same as discussed in paragraph 8-4D above.

F. Multiple Wedges/Multiple Intersecting Planes. Multiple wedges can be formed by the intersection of four or more sets of discontinuities (Figure 8-1F). Although conceptually possible, the sliding failure of a multiple wedge system rarely occurs because of the potential for kinematic constraint.

G. Single Block/Circular Slip Path. Single block sliding failures along circular slip paths

are commonly associated with soil slopes. However, circular slip failures may occur in highly weathered and decomposed rock masses, highly fractured rock masses, or in weak rock such as clay shales and poorly cemented sandstones.

8-5. TOPPLING FAILURE MODE.

Toppling failure involves overturning or rotation of rock layers. Closely spaced, steeply dipping discontinuity sets that dip away from the slope surface are necessary prerequisites for toppling. In the absence of cross jointing, each layer tends to bend downslope under its own weight, thus generating flexural cracks. If frequent cross joints are present, the layers can topple as rigid columns. In either case, toppling is usually initiated by layer separation with movement in the direction of the excavation. Layer separation may be rapid or gradual. Rapid separation is associated with block weight and/or stress relief forces. Gradual separation is usually associated with environmental processes such as freeze/thaw cycles.

8-6. SLOUGHING FAILURE MODE.

Sloughing failures are generally characterized by occasional rock falls or localized slumping of rocks degraded by weathering. Rock falls occur when rock blocks become loosened and isolated by weathering and erosion. Some rocks disintegrate into soil-like material when exposed to repeated wetting and drying cycles. This material can fail in a fashion similar to shallow slump type failures commonly associated with soil slopes. Both rock falls and localized slumping constitute more of a maintenance problem than a major slope instability threat. However, slopes in sedimentary rock that are interbedded with shale layers can experience major slope failures initiated by localized deterioration of the shale layers. Deterioration of the shale layers leads to the undermining and hence failure of the more competent overlying layers.

8-7. ADDITIONAL FACTORS INFLUENCING SLOPE STABILITY.

The geometric boundaries imposed by the orientation, spacing, and continuity of the joints, as well as the free surface boundaries imposed by the excavation, define the modes of potential failure. However, failure itself is frequently initiated by additional factors not related to geometry. These factors include erosion, ground water, temperature, in-situ stress, and earthquake-induced loading.

A. Erosion. Two aspects of erosion need to be considered. The first is large scale erosion, such as river erosion at the base of a cliff. The second is relatively localized erosion caused by groundwater or surface runoff. In the first type, erosion changes the geometry of the potentially unstable rock mass. The removal of material at the toe of a potential slide reduces the restraining force that may be stabilizing the slope. Localized erosion of joint filling material, or zones of weathered rock, can effectively decrease interlocking between adjacent rock blocks. The loss of interlocking can significantly reduce the rock mass shear strength. The resulting decrease in shear strength may allow a previously stable rock mass to move. In addition, localized erosion may also result in increased permeability and ground-water flow.

B. Ground Water. Ground water occupying the fractures within a rock mass can significantly reduce the stability of a rock slope. Water pressure acting within a discontinuity reduces the effective normal stress acting on the plane, thus reducing the shear strength along that plane. Water pressure within discontinuities that run roughly parallel to a slope face also increase the driving forces acting on the rock mass.

C. Temperature. Occasionally, the effects of temperature influence the performance of a rock slope. Large temperature changes can cause rock to spall due to the accompanying contraction and expansion. Water freezing in discontinuities causes more significant damage by loosening the rock mass. Repeated freeze/thaw cycles may result in gradual loss of strength. Except for periodic maintenance requirements, temperature effects are a surface phenomenon and are most likely of little concern for permanent slopes. However, in a few cases, surface deterioration could trigger slope instability on a larger scale.

D. State of Stress. In some locations, high in-situ stresses may be present within the rock mass. High horizontal stresses acting roughly perpendicular to a cut slope may cause blocks to move outward due to the stress relief provided by the cut. High horizontal stresses may also cause spalling of the surface of a cut slope. Stored stresses will most likely be relieved to some degree near the ground surface or perpendicular to nearby valley walls. For some deep cuts, it may be necessary to determine the state of stress within the rock mass and what effects these stresses may have on the cut slope.

SECTION II. METHODS FOR ASSESSING STABILITY

8-8. GENERAL.

This section presents a brief review of some of the more commonly used methods for assessing the stability of slopes cut in rock masses. The method selected for analyses depends on the potential failure mode and, to some extent, the preference of the District Office responsible for the analyses. In this respect, the discussions will be divided according to potential failure modes. The potential failure modes include sliding, toppling, and localized sloughing. A detailed discussion of each of the various methods is beyond the scope of this manual. Hoek and Bray (1974), Canada Centre for Mineral and Energy Technology (1977a), Kovari and Fritz (1989) and Hendron, Cording and Aiyers (1980) provide general discussions on analytical methods for accessing the stability of rock slopes. Specific references are given that provide in-depth details for each of the methods as they are discussed.

8-9. SLIDING STABILITY ANALYSES.

The majority of the methods used in analyzing the sliding stability of slopes cut into rock masses are based on the principles of limit equilibrium. The mathematical formulation of the various methods depends on the three general modes of sliding failure illustrated in Figure 8-1. These three general modes include planar slip surfaces, three-dimensional wedge-shaped slip surfaces, and circular slip surfaces. Since the majority of sliding stability problems are indeterminate, a number of assumptions must be made about the location, orientation, and possible magnitude of the forces involved in the analyses. Different methods are presented below along with a short description of the assumptions that are made, as well as the general procedure used for the analyses.

A. Planar Slip Surfaces. The analyses of planar slip surfaces assume that stability can be adequately evaluated from two-dimensional considerations. The following discussions summarize a number of different methods for analyzing the stability of planar slip surfaces. The methods are not all-inclusive but rather are representative of commonly used methods that are currently available.

(1) Simple plane method. The simple plane method is applicable to slopes in which the potential slip surface is defined by a single plane, as illustrated in Figure 8-1A. The method is based on equi-

librium between driving and resisting forces acting parallel and perpendicular to the potential slip surface. Mathematical expressions of the simple plane method can be found in most elementary physics text books. Convenient expressions are provided by Kovari and Fritz (1989).

(2) Two-dimensional wedge method. The two-dimensional wedge is suited for cases in which the potential failure surface of a rigid rock mass can be closely approximated by two or three planes. Hence, the method assumes that the potential failure mass can be divided into two or three two-dimensional wedges. A simplified approach assumes that forces between the wedges are horizontal. The horizontal force assumption generally results in a factor of safety that is within 15% (generally on the conservative side) of more accurate techniques which satisfy all conditions of equilibrium. Lambe and Whitman (1969) provide a detailed discussion and an example of the method.

(3) Generalized slip-surface methods for a rigid body. Generalized slip-surface methods refer to those methods that are used to solve two-dimensional rigid body stability problems using potential slip surfaces of any arbitrary shape. In this respect, the slip surfaces may be curvilinear in shape or defined by an assemblage of any number of linear segments as illustrated in Figure 8-1B. Of the available generalized slip-surface methods, the two best known methods were proposed by Janbu (1954) and Morgenstern and Price (1965).

a. Janbu's generalized slip-surface method is an iterative procedure using vertical slices and any shape slip-surface. The procedure, in its rigorous form, satisfies all conditions of equilibrium to include vertical and horizontal force equilibrium, moment equilibrium of the slices, and moment equilibrium of the entire slide mass. Complete equilibrium requires the solution of both shear and normal forces acting between slices. In the solution for the side forces, Janbu's method assumes the point of side force application as well as the line of action of all the side forces. Janbu (1973) provides a detailed discussion of theory and application.

b. Morgenstern and Price's generalized slip-surface method is similar to Janbu's method in that the procedure incorporates the interaction between a number of vertical slices. Complete equilibrium is achieved by assuming the values of variable ratios between the shear and normal forces acting on the sides of each slice. Morgenstern and Price (1965) provide a detailed discussion of the method.

(4) Generalized slip-surface methods for two or more rigid bodies. Generalized slip-surface meth-

ods for two or more rigid bodies refer to those analytical methods used to solve two-dimensional stability problems. In this special case, sliding can occur along the base of each body, as well as between each body, as illustrated in Figure 8-1C. At least three methods are available for analyzing this special case. These three methods include methods proposed by Kovari and Fritz (1989) and Sarma (1979), as well as the distinct element numerical model method (e.g., Cundall 1980).

a. Kovari and Fritz's (1989) method provides a relatively simple solution for the factor of safety of two or more adjacent blocks subject to sliding. The potential slide surface along the base of each block is represented by a single plane. Blocks are separated by planes of discontinuity that may be inclined at arbitrary angles with respect to the base of the potential slide plane. The method satisfies force equilibrium. Moment equilibrium is not considered. In this respect, solutions for the factor of safety tend to be conservative.

b. Sarma (1979) proposed a comprehensive solution to the two-dimensional, multiple-block sliding problem that satisfies both moment and force equilibrium. The method utilizes slices that can be nonvertical with nonparallel sides. Solution for the factor of safety requires an iterative process. As such, from a practical point of view, it is usually more convenient to program the method for use on programmable calculators or personal computers.

c. The distinct element method (e.g., Cundall 1980) is based on equations of motion for particles or blocks. The method offers a useful tool for examining the phenomenology and kinematics of potentially unstable slopes.

B. Three-dimensional Wedge-Shaped Slip Surfaces. The majority of potentially unstable rock slopes can be characterized as three-dimensional wedge problems, as illustrated in Figure 8-1D, 8-1E, and 8-1F. The analytical analysis of three-dimensional problems is substantially simplified if the geotechnical professional responsible for the stability analysis is conversant with the use of stereographic projection. Stereographic projection allows convenient visualization of the problem being analyzed as well as the definition of geometric parameters necessary for analysis. Goodman (1976), Hoek and Bray (1974), and Priest (1985) provide detailed discussions of theory and application of stereographic projection techniques. Once the problem geometry has been defined, an analytical method can be selected for assessing the sliding stability of the slope. For convenience of discussion, methods for assessing sliding stability will be divided into two categories: methods for single three-dimensional wedges and methods for multiple three-dimensional wedges.

(1) Three-dimensional single wedge methods. Three-dimensional single wedge methods are applicable to slopes in which the potential instability is defined by a single rigid wedge, as illustrated in Figures 8-1D and 8-1E. Sliding may occur along one or more planar surfaces. As a rule, analytical solutions for the factor of safety are based on the principles of limit equilibrium in which force equilibrium is satisfied. A large number of expressions for the solution of factors of safety are reported in the literature. Hendron, Cording, and Aiyer (1980), Hoek and Bray (1974), and Kovari and Fritz (1989) provide expressions and detailed discussions of the method. Hendron, Cording, and Aiyer (1980) and Chan and Einstein (1981) also provide methods for addressing potential block rotation as well as transverse sliding.

(2) Three-dimensional, multiple-wedge methods. Although conceptually possible, multiple three-dimensional wedge systems seldom fail in sliding because of the potential for kinematic constraint. Generalized analytical solutions for the factor of safety in such cases are not readily available. In this respect, three-dimensional distinct element methods (Cundall 1980) offer a means of evaluating the kinematics of potentially unstable slopes.

C. Circular Slip Surfaces. As in planar slip surfaces, the analyses of circular slip surfaces assume that stability can be adequately evaluated from two-dimensional considerations, as illustrated in Figure 8-1G. The methods are generally applicable to rock slopes excavated in weak intact rock or in highly fractured rock masses. Of the various circular slip surface methods available, two of the more commonly used include the ordinary method of slices and the simplified Bishop method.

(1) Ordinary method of slices. The ordinary method of slices (EM 1110-2-1902) is also known as the Swedish Circle Method or the Fellenious Method. In this method, the potential sliding mass is divided into a number of vertical slices. The resultant of the forces acting on the sides of the slices act parallel to the base of that particular slice. Only moment equilibrium is satisfied. In this respect, factors of safety calculated by this method are typically conservative. Factors of safety calculated for flat slopes and/or slopes with high pore pressures can be on the conservative side by as much as 60%, at least when compared with values from more exact solutions.

(2) Simplified Bishop method. The simplified Bishop method (Janbu, et al., 1956) is a modifica-

tion of a method originally proposed by Bishop (1955). In the simplified method, forces acting on the sides of any vertical slice are assumed to have a zero resultant in the vertical direction. Moment equilibrium about the center of the slip surface circle as well as force equilibrium is satisfied. There is no requirement for moment equilibrium of individual slices. However, factors of safety calculated with this method compare favorably with values obtained from more exact solution methods.

8-10. TOPPLING STABILITY ANALYSES.

Two-dimensional considerations indicate that toppling can occur if two conditions are present. In this respect, toppling can occur only if the projected resultant force (body weight plus any additional applied forces) acting on any block of rock in question falls outside the base of the block and the inclination of the surface on which the block rests is less than the friction angle between the block and surface. However, in actual three-dimensions, rock slopes consist of a number of interacting blocks that restrict individual block movement. As a result, the mechanism is likely to be a complex combination of sliding and toppling. Due to the complexities of failure, generalized analytical methods that attempt to solve for the factor of safety have not been developed. Three-dimensional numerical methods, such as the distinct element method, can, however, offer insight as to the kinematics of failure.

8-11. LOCALIZED SLOUGHING ANALYSES.

Localized sloughing failures refer to a variety of potential failure modes. These modes can range from rotational failure of individual blocks to minor sliding failures of individual small blocks or mass of rock. These types of potential instability are frequently treated as routine maintenance problems and, as such, are seldom analyzed for stability.

8-12. PHYSICAL MODELING TECHNIQUES.

In addition to the analytical methods, there exist a number of physical modeling techniques used for problems where analytical techniques may not be valid or may be too complex. Available methods include the Base Friction Model, Centrifuge Model, and small-scale models. All of these techniques have shortcomings in that basic parameters to include length, mass, and strength must be scaled. The difficulty arises in that all three parameters must be scaled in the same proportions. Simultaneous scaling requirements are difficult to achieve in practice. Therefore, it is common to scale the most important parameter(s) accurately and then attempt to relate the influence of the lesser important parameters to the test results. Physical modeling techniques are discussed by Hoek and Bray (1974) and Goodman (1976).

A. Base Friction Modeling. This modeling technique uses a frictional rolling base in the form of a long sheet or a conveyor-like belt that simulates gravity. The model material is typically a sand-flour-vegetable oil material that closely models friction angles of discontinuous rock. A two-dimensional model of the slope or excavation is formed on the table. As the belt moves, the model slowly deforms. The technique cannot be used to model dynamic loadings. It is an excellent method to investigate the kinematics of jointed two-dimensional systems.

B. Centrifuge Modeling. Centrifuge modeling attempts to realistically scale body forces (i.e., gravitational forces). In this respect, centrifuge modeling may be a possible solution in cases where gravity plays an important role. Centrifuge methods are presently expensive, and the available centrifuges typically have long waiting lists. Generally, these machines only allow rather small models to be evaluated. Also, instrumentation of these models is required as one cannot scrutinize the model during testing, except perhaps with the help of a visual aid.

C. Scaled Models. These models are straightforward; however, they require model materials to build the scale model. The model material development is difficult due to the previously mentioned scaling problems. Use of heavy materials, such as barite, might be of some use in scaling gravitational effects. In addition to scaling associated with modeling requirements, the scale effects associated with shear strength selection must be also be considered, as discussed in Chapter 4 of this manual.

8-13. DESIGN CONSIDERATIONS.

A rock slope is accessed to be stable or potentially unstable depending on the value of the calculated factor of safety. The calculated factor of safety is primarily dependent upon the geometry of the potential failure path selected for analyses and the shear strength representative of the potential failure surface. In addition, other factors, such as ground water conditions, potential for erosion, seis-

mic loading, and possible blast-induced loosening of the rock mass must also be considered.

A. Factor of Safety. For major rock slopes where the consequence of failure is severe, the minimum required calculated factor of safety is 2.0. For minor slopes, or temporary construction slopes where failure, should it occur, would not result in bodily harm or a major loss of property, the minimum required factor of safety is 1.3. The minimum required factor of safety for rock slopes subject to and assessed for seismic loading is 1.1. Any relaxation of these values will be allowed only with the approval of CECW-EG and should be justified by comprehensive studies of such a nature as to reduce uncertainties to a minimum.

B. Critical Potential Failure Paths. For a given rock slope, a number of potential failure paths are kinematically possible. Each kinematically possible failure path must be analyzed. The critical potential failure path is that potential slip surface which results in the lowest value for the factor of safety. For a rock slope to be judged safe with respect to failure, the factor of safety calculated for the critical potential failure path must be equal to or greater than the appropriate minimum required factor of safety.

C. Representative Shear Strength. Procedures for selecting appropriate shear strengths representative of potential failure paths are discussed in Chapter 4 of this manual.

D. Ground Water Conditions. Unlike natural rock slopes, cut slopes must be analyzed prior to excavation. Hence, while fluctuations in ground water levels may be known prior to design, the influence on these fluctuations due to excavation of a slope is difficult to predict. In this respect, assumptions pertaining to the phreatic surface and potential seepage pressures should be made on the conservative side.

E. Effects of Erosion. Certain argillaceous rock types (e.g., some shales) are susceptible to erosion caused by slaking upon repeated wetting and drying cycles. Soft sedimentary rocks, in general, are also susceptible to erosion processes due to normal weathering, stream flow, or wave action. In this respect, stability analyses must either account for the effects of potential erosion (i.e., loss of slope toe support and/or undermining of more competent upper layers) or the overall design must provide provision to control the effects of erosion.

F. Seismic Loading. Where applicable, the stability of rock slopes for earthquake-induced base motion should be checked by assuming that the specified horizontal and vertical earthquake accelerations act in the most unfavorable direction. In this respect, earthquake-induced forces acting on a potentially unstable rock mass may be determined by a quasi-static rigid body approach in which the forces are estimated by Equations 7-8 and 7-9, as given in Chapter 7 of this manual.

G. Potential Blast Effects. Shear strengths selected for design analyses are generally based on preconstruction rock mass conditions. Rock slopes are commonly excavated by drill and blast techniques. If improperly used, these excavation techniques can significantly alter the material properties of the rock mass comprising the slope. These alterations are more commonly evident as loosened rock, which results in a reduction of strength. Design analyses must either account for potential blast-induced loosening with subsequent loss of strength, or ensure that proper drill and blast procedures are used in the excavation process. Proper drill and blast procedures are given in EM 1110-2-3800.

SECTION III. TREATMENT METHODS AND PLANNING CONSIDERATIONS

8-14. GENERAL.

The stability assessment of rock slopes frequently indicates an impending failure is possible. In such cases, a number of methods are available for improving the overall stability. An appreciation of the mechanics associated with rock slope stability together with an understanding of treatment methods for improving the stability of potentially unstable slopes permit the detailed planning and implementation of a slope stability program.

8-15. TREATMENT METHODS.

The available treatment methods include alteration of slope geometry, dewatering to increase resisting shear strength, rock anchors, and toe berms protection to prevent slaking and erosion effects.

A. Slope Geometry. In the absence of an imposed load, the forces that tend to cause the instability of a slope are a direct function of both slope height and angle of inclination. A reduction of slope height and/or angle of inclination reduces the driving forces and, as a result, increases stability. In addition, since the majority of rock slope stability problems are three-dimensional in nature, a few degrees of rotation in the strike of the slope can, in some cases, cause a potentially unstable slope to become kinematically stable.

B. Dewatering. The presence of ground water within a rock slope can effectively reduce the normal stress acting on the potential failure plane. A reduction in normal stress causes a reduction in the normal stress dependent friction component of shear strength. Ground water induced uplift can be controlled by two methods, internal drains and external drains. In this respect, drainage is often the most economical and beneficial treatment method.

(1) Internal drains. Properly designed and installed internal drains can effectively reduce ground water levels within slopes, thereby increasing stability. The specific design of an effective drain system depends on the geohydraulic characteristics of the rock mass (i.e., joint spacing, condition, and orientation, as well as source of ground water). As a minimum, an effective drain system must be capable of draining the most critical potential failure surface. In climates where the ground surface temperature remains below freezing for extended periods of time, the drain outlet must be protected from becoming plugged with ice. Hoek and Bray (1974) describe various types of internal drains.

(2) External drains. External or surface drains are designed to collect surface runoff water and divert it away from the slope before it can seep into the rock mass. Surface drains usually consist of drainage ditches or surface berms. Unlined ditches should be steeply graded and well maintained.

C. Rock Anchors. Rock bolts, as well as grouted in place reinforcement steel and cables, are commonly used to apply restraining forces to potentially unstable rock slopes. Rock anchors may be tensioned or untensioned depending, primarily, on the experience and preference of the District Office in charge of design. It must be realized, however, that untensioned anchors rely on differential movement of the rock mass to supply the necessary resisting force and that very little cost is involved in tensioning. Where deformations must be minimized or where initial resisting forces must be assured, the tensioning of rock anchors upon installation may be required.

D. Erosion Protection. Shotcrete, frequently with the addition of wire mesh and/or fibers, is an effective surface treatment used to control slaking and raveling of certain argillaceous rock types that can lead to erosion problems. The treatment also prevents loosening of the rock mass due to weathering processes and provides surface restraint between rock bolts.

E. Toe Berms. Toe berms provide passive resistance that can be effective in improving the stability of slopes that the critical potential failure plane passes within close proximity to the toe of the slope.

PLANNING A SLOPE STABILITY PROGRAM

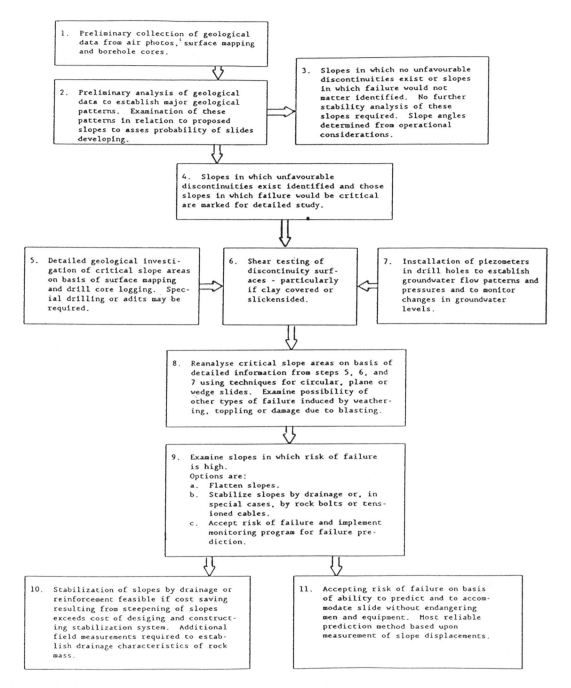

FIG. 8-2. Analysis of Stability of Slopes (Modified from Hoek and Bray 1981)

8-16. PLANNING CONSIDERATIONS.

With the design of numerous slopes or extremely long slopes, it is economically imperative that a system be followed which will eliminate naturally stable or noncritical slopes from study at a very early stage of investigation and allow concentration of effort and resources on those slopes that are critical. In this respect, a rock slope design flow chart which shows the steps required for design of rock slopes has been proposed by Hoek and Bray (1981) and is presented in Figure 8-2 with some modifications. The approach to the design of a slope is proposed in two phases.

A. Phase One. The first phase involves preliminary evaluations of available geologic data, which may include air photo interpretations, surface mapping, and gathering of data from rock cores from boreholes. Preliminary stability studies are then conducted using estimates of shear strengths of the discontinuities from index tests, experience, and back analyses of existing slope failures in the area. These preliminary studies should identify those slopes that are obviously stable and those in which there are some risks of failure. Slopes that are proven to be stable from the preliminary analysis can be designed on the basis of operational considerations.

B. Phase Two. Those slopes proven to have a risk of failure require further analyses based on more detailed information of geology, ground water, and mechanical properties of the rock mass. These analyses should consider the widest possible range of conditions that affect the stability of the slope. Slopes that are shown by detailed analyses to have an unacceptably high risk of failure must be redesigned to include stabilization measures. The operational and cost benefits of the stabilization measures should be compared with their implementation cost to determine the optimum methods of stabilization. The risk of failure for some slopes may be considered acceptable if slope monitoring would allow failures to be predicted in advance and if the consequences of a failure can be made acceptable.

CHAPTER 9

ANCHORAGE SYSTEMS

9-1. SCOPE.

This chapter provides guidance for the design and evaluation of anchor systems used to prevent the sliding and/or overturning of laterally loaded structures founded on rock masses. This chapter supplements guidance provided in EM 1110-1-2907. The chapter is divided into two sections: Modes of Anchor-Rock Interaction and Methods of Analyses.

SECTION I. MODES OF ANCHOR-ROCK INTERACTION

9-2. GENERAL.

Anchor systems may be divided into two general categories—tensioned and untensioned. The primary emphasis in the design, or selection of an anchorage system, should be placed on limiting probable modes of deformation that may lead to failure or unsatisfactory performance. The underlying premise of anchorage is that rock masses are generally quite strong if progressive failure along planes of low strength can be prevented. Both tensioned and untensioned anchors are suitable for the reduction of sliding failures in, or on, rock foundations. Tensioned anchor systems provide a means for prestressing all, or a portion, of a foundation, thus minimizing undesirable deformations or differential settlements. Preconsolidation of rock foundations results in joint closure and what appears as strain hardening in some foundations.

9-3. TENSIONED ANCHOR SYSTEMS.

A typical prestressed anchorage system is shown in Figure 9-1. The use of grouted anchorages is practically universal, particularly with high-capacity tendon systems. Upon tensioning, load is transferred from the tensioning element, through the grout, to the surrounding rock mass. A zone of compression is established (typically assumed as a cone) within the zone of influence. Tensioned anchor systems include rock bolts and rock anchors, or tendons. The following definitions are as given in EM 1110-1-2907.

A. Rock Bolt. A rock bolt is a tensioned reinforcement element consisting of a rod, a mechanical or grouted anchorage, and a plate and nut for tensioning or for retaining tension applied by direct pull or by torquing.

B. Prestressed Rock Anchor or Tendon. A prestressed rock anchor or tendon is a tensioned reinforcing element, generally of higher capacity than a rock bolt, consisting of a high strength steel tendon (made up of one or more wires, strands, or bars) fitted with a stressing anchorage at one end and a means permitting force transfer to the grout and rock at the other end.

9-4. UNTENSIONED ANCHOR SYSTEMS.

Untensioned rock anchors are generally referred to as rock dowels and are defined in EM

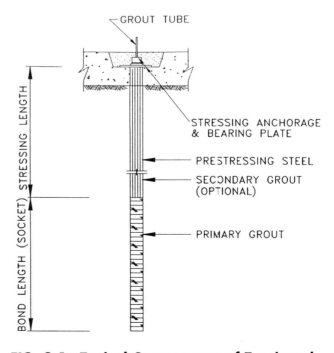

FIG. 9-1. Typical Components of Tensioned Rock Anchor (from EM 1110-1-2907)

1110-1-2907 as an untensioned reinforcement element consisting of a rod embedded in a mortar or grout-filled hole. Dowels provide positive resistance to dilation within a rock mass and along potentially unstable contact surfaces. In addition to the development of tensile forces resisting dilation, passive resistance against sliding is developed within a rock mass when lateral strains occur. The interaction between the dowel and the rock mass is provided through the cohesion and friction developed along the grout column, which bonds the rod and the rock. Untensioned anchor systems should not be used to stabilize gravity structures.

SECTION II. METHODS OF ANALYSIS

9-5. GENERAL.

Typically, analyses of systems used to anchor mass concrete structures consist of one of two methods: procedures based on classical theory of elasticity or procedures based on empirical rules or trial and error methods. The gap between the methods has been narrowed by research in recent years but has not significantly closed to allow purely theoretical analysis of anchor systems. The following discussions on methods of analyses are divided into tensioned and untensioned anchor systems.

9-6. ANALYSES FOR TENSION ANCHOR SYSTEMS.

The design and analysis of anchor systems include determination of anchor loads, spacing, depth, and bonding of the anchor. Safety factors are determined by consideration of the following failures: within the rock mass, between the rock and grout/anchor, between the grout and the tendon or rod, and yield of the tendon or top anchorage.

A. Anchor Loads. Anchor loads for prestressed tensioned anchors are determined from evaluation of safety factor requirements of structures. Anchors may be designed for stability considerations other than sliding to include overturning and uplift. Other factors must also be considered. However, anchor forces required for sliding stability assurance typically control design. Procedures for determining anchor forces necessary for stability of concrete gravity structures are covered in EM 1110-2-2200.

B. Anchor Depths. Anchor depths depend on the type of rock mass into which they are installed and the anchor pattern (i.e., single anchor, single row of anchors, or multiple rows of anchors).

The anchor depth is taken as the anchor length necessary to develop the anchor force required for stability. The entire anchor depth lies below the critical potential failure surface.

(1) Single anchors in competent rock. The depth of anchorage required for a single anchor in competent rock mass containing few joints may be computed by considering the shear strength of the rock mobilized around the surface area of a right circular cone with an apex angle of 90 deg. (see Figure 9-2A). If it is assumed that the in-situ stresses, as well as any stresses imposed on the foundation rock by the structure, are zero; then the shear strength can be conservatively estimated as equal to the rock mass cohesion. In such cases, the anchor depth can be estimated from Equation 9-1.

$$D = [(FS)\,(F)/c\,\pi]^{1/2} \qquad (9\text{-}1)$$

where

D = required depth of anchorage
FS = appropriate factor of safety
c = rock mass cohesion intercept; and
F = anchor force required for stability.

(2) Single row of anchors in competent rock. The depth of anchorage for a single row of anchors (Figure 9-2B) installed in competent rock and spaced a distance s apart may be computed as follows:

$$D = \frac{(FS)\,(F)}{cs} \qquad (9\text{-}2)$$

where F = anchorage force on each anchor. All other parameters are as previously defined.

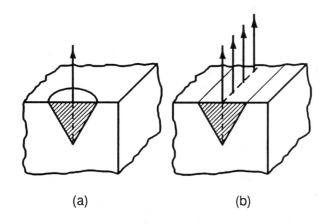

(a)   (b)

FIG. 9-2. Geometry of Rock Mass Assumed to Be Mobilized at Failure: (a) Individual Anchor in Isotropic Medium; (b) Line of Anchors in Isotropic Medium (after Littlejohn 1977)

(3) Multiple rows of anchors in competent rock. For a multiple row of anchors with rows spaced a distance l apart, typically, only the weight of the rock mass affected is used in calculations of resisting force. Under this assumption, the depth of anchorage required to resist a anchorage force F per anchor is computed as follows:

$$D = \frac{(FS)\,(F)}{\gamma\,l\,s} \qquad (9\text{-}3)$$

where γ = unit weight of the rock. All other parameters are as previously defined.

(4) Single anchor in fractured rock. In fractured rock, the strength of the rock mass subjected to a tensile force (the anchor force) cannot typically be relied upon to provide the necessary resistance. For this reason, only the weight of the affected one is considered. Based on this assumption, the depth of anchorage is completed as follows:

$$D = \left(\frac{3(FS)\,(F)}{\gamma\,\pi}\right)^{1/3} \qquad (9\text{-}4)$$

where γ = unit weight of the rock. All other parameters are as previously defined.

(5) Single row of anchors in fractured rock. As in the case of a single anchor in fractured rock, typically, only the weight of the affected wedge of rock is relied upon to provide the necessary resistance. Hence, for a single row of anchors in fractured rock spaced s distance apart, the anchorage depth is computed as follows:

$$D = \left(\frac{(FS)\,(F)}{\gamma\,s}\right)^{1/2} \qquad (9\text{-}5)$$

All other parameters are as previously defined.

(6) Multiple rows of anchors in fractured rock. For multiple rows of anchors with rows spaced l distance apart, again only the weight of the affected rock mass resists the anchor force. In this respect, Equation 9-3 is valid.

C. Anchor Bonding. The above equations, presented for analysis of anchor system, assume sufficient bond of the anchor to the rock such that failures occur within the rock mass. The use of grouted anchorages has become practically universal with most rock reinforcement systems. The design of grouted anchorages must, therefore, ensure against failure between the anchor and the grout, as well as between the grout and the rock. Experience and numerous pull-out tests have shown that the bond developed between the anchor and the grout is typically twice that developed between the grout

and the rock. Therefore, primary emphasis in design and analysis is placed upon the grout/rock interface. For straight-shafted, grouted anchors, the anchor force that can be developed depends upon the bond stress, described as follows:

$$F = \pi dL\tau \qquad (9\text{-}6a)$$

$$\tau = 0.5\tau_{ult} \qquad (9\text{-}6b)$$

where
d = effective diameter of the borehole;
L = length of the grouted portion of the anchor bond length (normally not less than 10 ft);
τ = working bond strength; and
τ_{ult} = ultimate bond strength at failure. Values of ultimate bond strength are normally determined from shear strength data, or field pull-out tests. In the absence of such tests, the ultimate bond stress is often taken as 1/10 of the uniaxial compressive strength of the rock or grout (whichever is less) (Littlejohn 1977) up to a maximum value of 4.2MPa (i.e., 600 psi).

9-7. DOWELS.

Structures should, in principle, be anchored, when required, to rock foundations with tensioned or prestressed anchorage. Since a displacement or partial shear failure is required to activate any resisting anchorage force, analysis of the contribution of dowels to stability is at best difficult. Dilation imparts a tensile force to dowels when displacements occur over asperities, but the phenomenon is rarely quantified for analytical purposes.

9-8. DESIGN CONSIDERATIONS.

A. Material Properties. The majorities of material properties required for the design of anchor systems are also typically required for the investigation of other aspects of the foundation design. The selection of appropriate material properties is discussed in Chapter 4 of this manual. Design anchor force derived from calculations not associated with sliding instability must consider the buoyant weight of rock where such rock is submerged below the surface water or ground water table. Tests not necessarily considered for typical foundation investigations but needed for anchor evaluations include rock anchor pull-out tests and chemical tests of

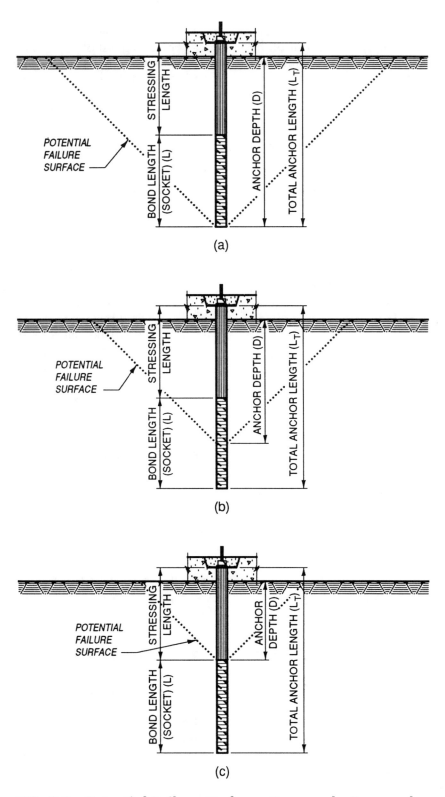

FIG. 9-3. Potential Failure Surfaces Commonly Assumed for Design of Anchor Depths in Rock Masses: (a) Potential Failure Initiates at Base of the Socket; (b) Potential Failure Initiates at Midpoint of the Socket; (c) Potential Failure Initiates at Top of the Socket

the ground water. Rock anchor pull-out tests (Rock Testing Handbook, RTH 323) provide valuable data for determining anchorage depth and anchor bond strength. Hence, a prudent design dictates that pull-out tests be performed in the rock mass representative of the foundation conditions and anticipated anchor depths. Ground water chemical tests establish sulphate and chloride contents to be used as a guide in designing the anchor grout mix. In addition, the overall corrosion hazard for the anchor tendon steel should be established by chemical analysis. Such analyses are used to determine the amount and type of corrosion protection required for a particular foundation.

B. Factors of Safety. The appropriate factor of safety to be used in the calculations of anchor force and anchorage depth must reflect the uncertainties and built-in conservatism associated with the calculation process. In this respect, anchor force calculations should be based on the factor of safety associated with sliding stability of gravity structures discussed in Chapter 7. Anchor age depth calculations based on the unit weight of the rock mass (Equations 9-3, 9-4, and 9-5) should use a minimum factor of safety of 1.5. All other anchorage depth calculations (i.e., Equations 9-1 and 9-2) should use a minimum factor of safety of 4.0 unless relaxed by CECW-EG for special circumstances.

C. Total Anchor Length. In addition to the anchor depth and anchor bonding considerations given by Equations 9-1 through 9-5 and Equation 9-6, respectively, the total anchor length (L_T) is controlled by the location at which the rock mass is assumed to initiate failure, should a general rock mass failure occur. Littlejohn and Bruce (1975) summarize the assumed location of failure initiation commonly used in practice. As indicated in Figure 9-3, three locations are commonly assumed: potential failure initiates at the base of the socket; potential failure initiates at the midpoint of the socket; or potential failure initiates at the top of the socket. The implication with respect to the total anchor length imposed by each failure location assumption is as shown in Figure 9-3. For the design of anchors in competent or fractured rock masses where the bond length is supported by pull-out tests, the potential for rock mass failure is assumed to initiate at the base of the anchor as shown in Figure 6-3A. For preliminary design where pull-out tests are not yet available or in highly fractured and very weak material, such as clay shale, the potential for

failure is assumed to initiate at the midpoint of the socket as shown in Figure 6-3B. However, in the case of highly fractured and very weak material, pull-out tests must be performed to verify that the bond length is sufficient to develop the ultimate design load as specified in EM 1110-2-2000. Any relaxation in total anchor length requirements must be approved by CECW-EG.

D. Corrosion Protection. The current industry standard for post-tensioned anchors in structures requires double corrosion protection for all permanent anchors.

E. Design Process. The rock anchor design process is conveniently divided into two phases: the initial design phases and the final detailed phase. Additional details are provided in EM 1110-2-2200 and Post-Tensioning Institute (1986).

(1) Initial phase. The design process is initiated by an evaluation which finds that a given structure is potentially unstable without additional restraining forces. If the potential instability is due to potential for sliding, the magnitude of restraining forces is calculated according to procedures given in EM 1110-2-2200. Restraining forces necessary to control other modes of potential instability, such as overturning, uplift pressures, or excessive differential deformations, are determined on a case-by-case basis. The magnitude of the required restraining force is evaluated with respect to the economics and practicality of using rock anchors to develop the necessary force.

(2) Final phase. The final detailed design phase is a trial and error process that balances economic and safety considerations with physical consideration of how to distribute the required restraining force to the structure and still be compatible with structure geometry and foundation conditions. While sequential design steps reflect the preference of the District Office, general design constraints usually dictate that the total restraining force be divided among a number of anchors. The number of anchors and, hence, the spacing between anchors and anchor rows, as well as the anchor orientation and installation details, are controlled by the geometry of the structure. Foundation conditions control the anchorage depth as well as the amount and type of corrosion protection. Anchor depths between adjacent anchors should be varied to minimize adverse stress concentrations.

CHAPTER 10

INSTRUMENTATION

10-1. SCOPE.

This chapter provides general guidance for the selection and use of instrumentation to monitor cut slopes, such as might be necessary for the construction of rock foundations and roads as well as structures founded on rock such as dams, lock walls, and retaining structures. Instrumentation for monitoring ground vibrations, water levels, and pore-water pressure measurements are discussed in more detail than other instrumentation because of their widespread use. The limitations as well as data interpretation and evaluation considerations are also discussed. Detailed descriptions and installation considerations of the various types of instrumentation discussed herein can be found in the referenced publications. This chapter is divided into four sections as follows: Planning Considerations, Typical Applications, Types of Instruments, and Data Interpretation and Evaluation.

SECTION I. PLANNING CONSIDERATIONS

10-2. GENERAL.

Instrumentation is necessary on a project to (1) ensure that design criteria are being met, thereby assuring the safety of the structure, (2) gain information valuable to future project design, (3) monitor suspected problem areas to determine safety and remedial measures required, and (4) monitor effectiveness of remedial measures.

10-3. PROGRAM INITIATION.

An instrumentation program should be planned during the design of a project. The specific areas and phases of the project from which data need to be gathered are determined using the rock mechanics analyses and models discussed in previous chapters. In order to obtain the most complete picture of how a rock mass is responding to the construction and operation of a project, instrumentation should be installed where possible before or during construction. Early installation rarely in-

creases the cost of the instrumentation program, but does require more planning.

10-4. COST CONTROL.

The instrumentation program should be well planned to ensure that all necessary data will be collected and that excessive costs are not incurred. The main expenses of an instrumentation program include instrument purchase, installation, maintenance, data gathering, and data interpretation. Excessive costs in each of these areas are incurred if instrument types and placement are planned unwisely, leading to more instrumentation than is necessary for the intended purpose or difficulty in interpreting data due to lack of information. The instrumentation program must be flexible enough to allow for changes necessary due to actual conditions encountered during construction.

10-5. TYPES AND NUMBER OF INSTRUMENTS.

The parameters that are most often measured are deformation, load/stress, pore-water pressures and water levels, and ground vibrations. The types of instrumentation used to measure these parameters are listed in Table 10-1. The number of instruments and various types that will be required on a specific project are dependent on the purpose of the structure and the geologic conditions. The

TABLE 10-1. Types of Rock Foundation Instruments

Deformation (1)	Load/stress (2)	Pore-water pressure (3)	Ground vibration (4)
Surveying Inclinometers Extensometers Settlement Indicators Heave points	Load cells Piezometers Uplift pressure cells	Piezometers	Seismographs

instrumentation program for every project should be designed specifically for that project and the expected conditions and should use the principles of rock mechanics. Rock instrumentation must reflect conditions over a large area of rock. Measurements made over small areas will yield data so influenced by small random features that it will be meaningless. Great care should be taken to ensure that the particular instrumentation used will yield the type of information required at the necessary accuracy. An instrumentation program should be kept as simple as possible and still meet the objectives of the program. A complicated instrument is generally harder to maintain and less reliable than a simple type. Simple, direct measurements are most easily and quickly interpreted.

SECTION II. APPLICATIONS

10-6. GENERAL.

This section describes some of the more common applications of rock mechanics instrumentation. The discussions are divided into two general topic areas related to project features addressed in this manual. These two topic areas include cut slope instrumentation and structure/foundation instrumentation.

10-7. CUT SLOPE INSTRUMENTATION.

The number, types, and location of instruments used in cut slopes are highly dependent on the cut configuration, the geologic conditions that are involved, and the consequence should a failure occur. As a rule, however, instrumentation associated with cut slopes can be grouped into instruments used to make surface measurements and those used to make subsurface measurements.

A. Surface Measurements. Surface measurement instruments are primarily used to measure surface deformations. Since surface instrumentation reveals little as to underlying mechanisms causing deformation, the instrumentation is used to detect new areas of distress or precursor monitoring of rock masses subject to impending failure. The degree of precision required by the intended purpose of instrumentation dictates the type of instrument used to measure deformation.

(1) Surveying. If the slope is stable, then periodic surveying of the floor and sidewalls using permanent monuments and targets may be the only instru-

mentation required. Precise, repetitive surveying of a network of such survey points is a relatively inexpensive method of detecting slope movement, both vertical and lateral. When a problem is detected, surveying can be used to define the area of movement. Evaluation of problem areas is required to determine if additional instrumentation is required. Depending on other factors, surveying may be continued, perhaps with increasing frequency, until remedial measures appear to be inevitable. In other cases, the failure of the slope may be more acceptable than the cost of the remedial measures and surveying would be continued until the slope failed, to ensure the safety of personnel and equipment when failure occurs. Details of the instruments and surveying methods used may be found in TM 5-232, "Elements of Surveying," and TM 5-235, "Special Surveys."

(2) Surface deformation. In most cases, however, additional instrumentation will be required to provide the information that enables the investigator to find or to define the causes of the movement and to monitor the rate of movement. Tension cracks that appear at the crest of a slope or cut face may be monitored by surface type extensometers. This type of extensometer generally consists of anchor points installed on either side of the zone to be monitored. The zone may be one joint or crack or several such features. A tape or bar, usually composed of invar steel, is installed between the anchor points. A Newcastle extensometer may be installed on the tape to allow for very accurate readings that are necessary to measure the small initial indications of movement. For measuring larger movements, which would occur later and when continuous measurements are required, a bar and linear potentiometer can be installed between the stakes. (See Chapter 8 of the Canada Centre for Mineral and Energy Technology (1977b) for details.) If very large measurements are expected, a simple inexpensive system that uses a calibrated tape to measure the change in distance between the two anchor points should be used. The tape can be removed after a reading is made. This instrument aids in the determination of the surface displacement of individual blocks and differential displacements within an unstable zone. Dunnicliff (1988) provides an excellent review of the various types of surface-monitored extensometers.

B. Subsurface Measurements. Subsurface instrumentation provides greater detail of mechanisms causing distress. Because subsurface instruments require installation within a borehole and the cost associated with such installations, their use is typically limited to monitoring known features of potential instability or to investigate suspected features. Subsurface deformation measurements monitor the relative movement of zones of rock with respect to each other. Piezometric pressure measurement along zones of potential instability monitor the influence of ground water with respect to stability.

(1) Subsurface deformations. Subsurface deformations within rock slopes are commonly measured with one of two types of downhole instruments, inclinometers, or borehole extensometers.

a. Inclinometers are installed behind the slope, on flat slopes where drilling access is available, or into the slope and are bottomed in sound, stable rock. Successive measurements of deflections in the inclinometer are used to determine the depth, magnitude, and rate of lateral movement in the rock mass. While commonly installed in vertical boreholes, inclinometers are available that allow installation in inclined to horizontal boreholes. Because successive deflection measurements can be made at small intervals, the device is ideally suited to precisely locate and define as well as monitor zones of instability. Detailed descriptions of inclinometers can be found in EM 1110-2-1908 (Part 2), "Instrumentation of Earth and Rock Fill Dams," and Dunnicliff (1988).

b. Borehole extensometers are often placed into the face of a cut or slope to help in determining the zones behind the face that are moving. When a deep cut is being made, extensometers may be installed in the walls as the excavation progresses to monitor the response of the slope to an increasing excavation depth. Multiposition borehole extensometers (MPBX), rod or wire, are able to monitor relative movement of a number of different zones at varying distances behind the cut face. Such measurements help to determine which zones are potentially critical and the rate of movement. MPBX's are particularly helpful in distinguishing between surficial and deep-seated movement. Extensometers may be equipped with switches that automatically close and activate warning devices when a preset movement limit is reached. Unless care is taken to isolate downhole wires or rods, installations at great depths are not always practical due to the difficulty of obtaining a straight borehole. It is necessary to eliminate, as much as possible, the friction effects between the extensometer wire or rod and the borehole wall. Friction effects can introduce large errors that make interpretation of the data impossible. The maximum measurable deformation is relatively small, ranging from approximately 0.5 to several inches, but this limit can be extended by resetting the instrument. Extensometers are described in EM 1110-2-1908, Part 2, and Dunnicliff (1988).

(2) Piezometric pressure. Drainage of a cut slope is often necessary to increase its stability by reducing pore-water pressures in the slope. The effectiveness of any drainage measure should be monitored by piezometers. Piezometer data should also be used to determine when maintenance of a drainage system is necessary. Piezometers should be installed during site investigation activities to determine the ground-water system. Preconstruction installation is important not only for design of the project but also to determine if construction will adversely affect nearby ground-water users. Data should be obtained before, during, and after construction so that a cause-affect trend can be determined, if there is one. This information is very important if there are claims that conditions in nearby areas have been changed due to activities at the project. Piezometers are discussed in Section III of this chapter.

(3) Anchor loads. When the instruments discussed above indicate that remedial measures, such as rockbolts, are necessary to stabilize a slope, then these same instruments are used to monitor the effectiveness of the remedial measures. The actual load or tension acting on a rockbolt is monitored with a load cell. This information is to assure that bolts are acting as designed and that the maximum load on the bolt is not exceeded. A representative number of bolts in a system are usually monitored. The types of load cells include the hydraulic, mechanical, strain gaged, vibrating wire, and photoelastic. The strain-gaged load cell is the type most often used to monitor rockbolt systems. Load cells are described in the Rock Testing Handbook as well as Dunnicliff (1988).

10-8. FOUNDATION/STRUCTURE INSTRUMENTATION.

As in the case of cut slopes, foundations and structures, such as dams, lock walls, and retaining structures, may require a large number and variety of instruments. These instruments are frequently similar or the same as those required for slope monitoring and are divided into three general categories dependent on what observation is being measured. The three categories include deformation measurements, piezometric pressure measurements, and load/stress measurements.

A. Deformation Measurements. Deformations of foundations and structures are generally observed as apparent translation, rotation, or settlement/heave. Apparent deformations may actually be the result of a combination of the above deformation modes.

(1) Translation. Translation deformations caused by foundation/structure interactions are generally apparent as sliding along planes of weakness.

It is essential to define the planes along which translation occurs and evaluate the severity of the problem at an early stage. Translation measurements of foundations and structures are generally monitored with subsurface techniques discussed under cut slope instrumentation.

(2) Rotation. A tiltmeter may be used to determine the rate, direction, and magnitude of angular deformation which a rock mass, a structure, or a particular block of rock is undergoing. A tiltmeter, unlike an inclinometer, measures only at a discrete, accessible point. The device may be permanently buried with a remote readout or may be installed directly on the rock or structure surface. If there is weathered rock at the surface, the device may be mounted on a monument that is founded in or on intact rock. The tiltmeter consists of a reference plate, which is attached to the surface that is being monitored, and a sensing device. A portable sensing device may be installed on the reference plate for each reading, or a permanent, waterproof housing containing the sensing device may be installed directly on the surface to be monitored. In the second case, readings may be made from a remote readout station. Tiltmeters may also be installed directly on a structure. Tiltmeters are described in more detail in the Rock Testing Handbook and Dunnicliff (1988).

(3) Settlement/Heave. Settlement refers to compression of the foundation material whereas heave refers to expansion. Mechanisms that cause settlement are discussed in Chapter 5. Mechanisms that cause heave were also briefly discussed in Chapter 5, but are discussed in greater detail in Chapter 12.

a. Settlement of a foundation beneath a structure may be determined by repeated surveying of the elevation of a settlement gauge monument installed directly on the foundation and protected from frost and vandalism. Points on the structure itself may likewise be surveyed to determine settlement, especially if direct access to the foundation is not possible. Settlement indicators may also be used to measure settlement. Settlement indicators are capable of measuring single or multiple points and operate on the same principle as a manometer. In areas beneath buildings, or other areas where direct access to the instrument is not available, a remotely read instrument may be used as described by Hanna (1973). The instrument is installed in the foundation before the structure is built. The elevation of the measuring point is calculated using the elevation of the readout point and a pressure reading at the measurement point. The original elevation of the measuring point must be determined for comparisons to later readings.

b. The floor of an excavation may require monitoring for heave or rebound. Heave is not common in all rock or foundation conditions. Heave measurements give valuable information for use in design of other structures in similar rock masses and conditions. These measurements are also important to correlate performance with design assumptions, especially when the foundation is to support precise industrial or scientific equipment where little departure from the design criteria can be tolerated. Heave points are the most common technique used to measure rebound during excavation. Heave points usually consist of an anchor point that is placed in a borehole at or below the expected elevation at the bottom of the excavation. The elevation of the anchor is determined. The drill hole is filled with a bentonite slurry that contains a dye to aid in relocating the instrument hole during construction. As excavation proceeds, a probe of known length is lowered to the top of the anchor point and the elevation of the anchor point is determined by optical leveling. An alternative method uses a linear potentiometer as the sensing element in the borehole. This type of settlement gauge is described by Hanna (1973). Settlement/heave gauges are also described in EM 1110-2-1908 (Part 2) and Dunnicliff (1988). The method used for anchoring the reference point to the rock and protection of the instrument during construction are important considerations.

B. Piezometer Pressure Measurements. As in rock slopes, piezometers are often installed during site investigations and monitored to determine preconstruction conditions. A thorough understanding of the preconstruction conditions is very important not only for determining the effects of such conditions, especially seasonal variations, on the construction and operation of the structure, but also for determining the effects of the structure on the ground-water flow system. Dewatering activities, construction of ground-water cutoffs, and reservoir filling may affect local ground-water elevations and flow systems at some distance away from the project, possibly producing adverse affects. Once construction begins, piezometers that are not destroyed should continue to be monitored. This information can be used as an indication of how ground-water conditions and pore pressures change due to various construction activities such as removal of overburden or the added weight of the structure. Additional piezometers are installed when the structure is finished to monitor the performance of cutoffs and drainage systems as well as to measure pressures in the foundation underneath a structure or in abutments. The flow rate through the drainage system should be measured as another method of monitoring its performance. Unexplained changes in seepage rates may warn of a serious problem even before it is reflected by piezometer or other instrumentation data. Calibrated weirs or simply a stopwatch and calibrated container for lower flows are commonly used to measure drain flows. Other critical areas should also be instrumented as determined during design. Piezometers are described in more detail in Section III of this chapter.

C. Load/Stress Measurements. Instrumentation is frequently required to check design assumptions relating to stress distributions caused by rock/structure interactions, as well as to monitor zones of potential distress. Measurements of stress change in a foundation are made with earth pressure cells which may be installed at the interface of the structure and the rock or in a machined slot within the rock mass. Three commonly used pressure cells, to include vibrating wire, hydraulic (Gloetzl) and WES (similar to Carlson stress meter) type cells, are discussed in EM 1110-2-4300. It is necessary to install a piezometer near a pressure cell to isolate earth pressure changes from pore-water pressure changes. Pressure cells must be installed carefully to eliminate error caused by small localized stress concentrations.

D. Combined Measurements. As discussed in Chapter 5, settlement or heave is not frequently uniformly distributed across the foundation. In such cases, it may be necessary to monitor the effects of both settlement/heave and structural rotation. Instruments capable of monitoring these combined effects include plumb lines, inverted plumb lines, and optical plummets. These devices are thoroughly discussed in EM 1110-2-4300.

SECTION III. TYPES OF INSTRUMENTS AND LIMITATIONS

10-9. GENERAL.

Section II discussed the general application of a number of different types of instruments commonly used to monitor the performance of cut slopes and foundation rock/structure interactions. References were given that provided detailed descriptions, installation procedures, and limitations as well as advantages and disadvantages of various devices. This section will address two specific types of instruments: piezometers and ground motion/vibration monitoring devices. Piezometers have been mentioned previously but will be covered in greater detail here. Ground motion devices, considered to be location/site specific devices, will be briefly discussed in this section.

10-10. PIEZOMETERS.

Piezometers are used to measure pore-water pressures and water levels in the natural ground, foundations, embankments, and slopes. Piezometers are also used to monitor the performance of seepage control measures and drainage systems and to monitor the effect of construction and operation of the project on the ground-water system in the vicinity of the project. There are three basic types of piezometers: open-system (open standpipe), closed-system (hydraulic), and diaphragm (pneumatic and electrical, e.g., vibrating wire). The operation, installation, and construction of these piezometers are covered in detail in EM 1110-2-1908. The basic criteria for selecting piezometer types are reliability, simplicity, ruggedness, and life expectancy. Other considerations are sensitivity, ease of installation, cost, and the capability of being monitored from a remote observation point. Sometimes two or more types of piezometers may be required to obtain the most meaningful information at a particular site. One of the most important factors to be considered is the impact of hydrostatic time lag on the intended use of the piezometer data. Table 10-2 compares the different types of piezometers.

A. Open-system Piezometers. Open-system piezometers are the simplest types of piezometers, but they are also subject to the greatest hydrostatic time lag. They are best used in areas where slow changes in pore-water pressure are expected and the permeability is greater than 10^{-5} cm/sec. If rapid pore water pressure changes are expected, then open-system piezometers should only be used if the permeability is greater than 10^{-3} cm/sec (EM 1110-2-1908, Part 1 of 2).

B. Closed-system Piezometers. The rate of pore-water pressure changes has little effect on the measurements obtained with this type of piezometer. This type is commonly used to measure pore pressures during construction of embankments. The readout can be directed to a central location so that there is little interference with construction. However, the device must be checked often for leakage and the presence of air. Open-system piezometers should be installed near key closed-system piezometers to provide a check on the operation of the closed-system piezometer.

TABLE 10-2. Comparison of Piezometer Types

Basic type (1)	Relative volume demand (2)	Readout equipment (3)	Advantages (4)	Disadvantages (5)
Open-system (standpipe)	High	Water level finder	Simple; comparatively inexpensive; generally not subject to freezing; relatively long life; fairly easy to install; long history of effective operation.	Long time lag in most rock types; cannot measure negative pore pressure; cannot be used in areas subject to inundation unless offset standpipe used; must be guarded during construction; no central observation station is possible; requires sounding probe. Must be straight; difficulties possible in small diameter tubes if water levels significantly below 100 ft, or dip less than 45 deg.
Closed-system (hydraulic)	Medium to low	Usually bourdon gauge or man-ometer	Small time lag; can measure negative pore pressures; can be used in areas subject to inundation; comparatively little interference with construction; can be read at central observation stations.	Observation station must be protected against freezing; fairly difficult to install; fairly expensive compared to open systems; sometimes difficult to maintain an air-free system; most types are fragile; some types have limited service behavior records; requires readout location not significantly above lowest water level.
Diaphragm	Low to negligible	Specialized pressure transmitter or electronic readout	Simple to operate; elevation of observation station is independent of elevation of piezometer tip; no protection against freezing required; no de-airing required; very small time lag.	Limited performance data, some unsatisfactory experience; some makes are expensive and require expensive readout devices; fragile and requires careful handling during installation.
			Pneumatic. Electrical source not required; tip and readout devices are less expensive than for electrical diaphragm types.	Often difficult to detect when escape of gas starts; negative pressures cannot be measured; condensation of moisture occurs in cell unless dry gas is used; requires careful application of gas pressure during observation to avoid damage to cell.
			Electrical. Negative pressures can be measured; ideal for remote monitoring.	Devices subject to full and partial short-circuits and repairs to conductors introduce errors; some makes require temperature compensation and have problems with zero drift to strain gauges; resistance and stray currents in long conductors are a problem in some makes; zero drift possible.

1. Modified from Pit Slope Manual, Chapter 4, 1977 and EM 1110-2-1908 (Part 1).

C. Diaphragm Piezometers. Diaphragm piezometers can be used in the same situations as open and closed system piezometers. They are very sensitive to pore-water pressure changes, and the elevation difference between the piezometer tip and the readout point is not a limiting factor. The electrical diaphragm piezometer is complex and may be subject to instrument "zero" drift after calibration and installation, short circuits in the lead cable, stretch and temperature effects in long lead cables, and stray electrical currents.

10-11. GROUND MOTIONS/VIBRATIONS.

Ground motions/vibrations that can affect a rock foundation may be caused by earthquakes or blasting. Controlled blasting techniques, as discussed in Chapter 11, are used to minimize damage to foundations and adjacent structures caused by blasting. Seismographs should be used to monitor the levels of vibration actually being produced. Seismograph records (seismograms) are also used to provide a record of vibrations to ensure maximum levels are not exceeded that could cause damage to adjacent structures. Seismograph is a general term that covers all types of seismic instruments that produce a permanent record of earth motion. The three main types of seismographs measure particle displacement, velocity, and acceleration. The instruments used in different applications are discussed below.

A. Earthquakes. Measurement of earthquake motion assists in damage assessment after a significant earthquake and is necessary for improving the design of structures, especially dams, to better resist earthquakes. Guidance is given in EM 1110-2-1908 for determining which structures require instrumentation. The strong motion accelerograph and peak recording accelerograph are the principal instruments used to record earthquake motions on engineering projects, such as dams. The accelerograph measures particle acceleration in any direction or directions desired. The strong motion instruments generally record seismic motion between 0.01 g and 1.0 g. They are triggered by the minimum level of motion and record continuously during any motion above a preset minimum level and for a short time after motion ceases. The peak accelerograph records only the high amplitudes of the acceleration and does not make a continuous recording. This low-cost instrument is used only to supplement data from other accelerographs. One or two strong motion accelerographs may be located on a project, and several peak accelerographs may be located in other areas to obtain an idea of how the acceleration differs across the site. EM 1110-2-1908 provides additional discussions.

B. Blasting. As discussed in Chapter 11, construction blasting should be controlled to reduce damage by ground vibrations to the foundation being excavated and to nearby structures. Seismographs are used to monitor the ground vibrations caused by blasting. The peak particle velocity is normally used as an indication of potential damage; therefore, a velocity seismograph is normally used in engineering applications. The particle velocity can be inferred from the information obtained by other types of seismographs, but it is preferred to measure it directly so that an immediate record is available without extensive processing. EM 1110-2-3800, the Blaster's Handbook (Dupont de Nemours and Company 1977), and Dowding (1985) provide additional instrument descriptions.

10-12. LIMITATIONS.

There are certain requirements by which all types of field instrumentation should be evaluated. These include the range, sensitivity, repeatability, accuracy, and survivability of the instrument. The range must be adequate to measure the expected changes but not so great that sensitivity is lost. It is not always possible to accurately predict the magnitude of loads and deformations to be expected before construction. The most important of these factors may be repeatability, because this factor determines the quality of the data. The sensitivity required will vary with the application. Good sensitivity is required for early detection of hazards but may mean a reduction in the range and stability of the instrument. If an instrument with too narrow a range is chosen, all the necessary data may not be obtained. If an instrument with too large a range is chosen, then it may not be sensitive enough. Accuracy is difficult to define and to demonstrate. The anisotropy of a parameter must be predictable if the accuracy is to be determined. Calibration, consistency, and repeatability are also used in determining accuracy. The instrument chosen for a particular application must also be able to survive the often severe conditions under which it will be used. Cost should also be considered, and the least expensive way of obtaining good quality information should be used. Table 10-3 provides a summary of some of the major limitations of the various types of instrumentation that have been discussed. Ranges and sensitivities for different instrument types may vary between manufacturers and may change rapidly due to research and development, and so are not listed in this table. Many of the instruments are also easily modified by a qualified laboratory to meet the requirements of a particular job.

SECTION IV. DATA INTERPRETATION AND EVALUATION

10-13. READING FREQUENCY.

The frequency at which instrument readings are taken should be based on many factors and will vary by project, instrument type, availability of gov-

TABLE 10-3. Limitations of Rock Instrumentation

Instrument (1)	Measured parameter (2)	Limitations (3)
Inclinometer	Deformation	Life may be limited in hard rock due to sharp edges. Significant drilling costs.
Tiltmeter	Deformation	Measures one, near-surface discrete point. Subject to damage during construction. Difficult to detect spurious data. Must be protected from the environment. Subject to errors caused by bonding material.
Extensometers	Deformation	
Bar		Does not distinguish between deep-seated and surficial movement. Limited accuracy due to sag. Measures only one point. Significant drilling costs, a new drill hole required for each detection point.
Single point		
Multipoint rod		Limited to approximately 50-ft depth if each rod is not individually cased within the instrument hole. Experienced personnel should install them. May be damaged by borehole debris unless protected. Spring anchors may experience variable spring tension due to rock movement.
Multipoint wire		
Settlement indicators	Deformation	Hydraulic types require de-aired water. Corrections for temperature and barometric pressure differences are required. Access to drill collar is required for some types.
Heave point	Deformation	Accuracy is limited by surveying techniques used.
Load cells	Stress, load	
Hydraulic		Large size, poor load resolution, temperature sensitivity.
Mechanical		Nonlinear calibration curves.
Strain gauge		Requires waterproofing, long term stable bonding method and periodic recalibration.
Vibrating wire		Large size, expensive, poor temperature compensation, complicated readout, vulnerable to shock.
Photoelastic		Coarse calibration. Requires access to borehole collar.
Piezometers	Load, stress	See Table 10-2.
Uplift cells	Deformation	
Standpipe		Readings may require either of two methods, sounder or pressure gauge.
Diaphragm		Susceptible to damage during installation

ernment personnel to take readings, and location, and may even vary through time. The availability of government personnel to take the readings should be determined during the preparation of plans and specifications. If government personnel will not be available, provisions should be made to have this task performed by the construction contractor or by an A-E contractor. Some of the factors that should be evaluated include outside influences such as construction activities, environmental factors (rainfall events, etc.), the complexity of the geology, rate of ground movements, etc. Several sets of readings should be taken initially to establish a baseline against which other readings are to be evaluated. Daily or even more frequent readings may be necessary during certain construction activities, such as fill

placement or blasting. The rate of change of the condition that is being monitored may vary over time, dictating a change in the established frequency at which readings are taken. For example, an unstable slope may move slowly at first, requiring infrequent readings on a regular basis until a near failure condition is reached, at which time readings would have to be taken much more frequently. Readings of different types of instruments should be made at the same time. Concurrent readings enable the interpreter to take into account all the factors that might impact individual readings of specific parameters. For example, an increase in pore water pressure might coincide with increased slope movement. Standard forms should be used to record data when available or, if not, forms should be developed for specific instruments. Some forms are shown in EM 1110-2-1908. If possible, data should be reduced in the field and compared with previous readings so that questionable readings can be checked immediately. When large amounts of data must be managed, automatic recording devices that record data as printed output or on magnetic tape for processing by computer should be considered. Too many readings are not necessarily better than too few. An excess of data tends to bog down the interpretation process. A thorough evaluation of the purpose of the instrument program must be used to determine the optimum rate at which readings should be taken, thus ensuring that data are obtained when it is needed.

10-14. AUTOMATIC DATA ACQUISITION SYSTEMS.

Automatic data acquisition systems and computer data processing are very popular for obtaining and processing instrumentation data. Computer programs are available for reducing and plotting most types of data. Some of the advantages and disadvantages of these systems are given by Dunnicliff (1988). Use of computer processing can speed much tedious processing but should not replace examination of all of the data by an experienced person.

10-15. DATA PRESENTATION.

Most types of data are best presented in graphical form. Graphical presentation facilitates the interpretation of relationships and trends in the data. Readings are compared over time and with other instrument readings as well as with construction activities and changing environmental conditions. Observed trends should be compared with predicted trends to make an assessment of overall performance. The data should be displayed properly, or significant trends may be obscured or may become misleading. A thorough knowledge and understanding of the instrumentation, as well as some trial and error, is required to successfully accomplish good data presentation. Cookbook interpretation methods are available for some types of data such as that from inclinometers. Cookbook interpretation

is discouraged. Every instrument should be carefully and impartially analyzed by experienced personnel, taking all the available information into consideration.

10-16. DATA EVALUATION.

Factors to consider when evaluating instrumentation data include instrument drift, cross sensitivity, calibration, and environmental factors, such as temperature and barometric pressure. Instrument drift is the change in instrument readings over time when other factors remain constant. Drift can be caused by temperature fluctuations, power supply instability (weak battery), etc. If drift is not detected, it can lead to erroneous data interpretation. Periodic calibration of instruments, when possible, can reduce drift problems. Making repetitious readings also helps to detect and account for drift errors. Field calibration units may be available for some instrument types, such as inclinometers. Most instrumentation can be isolated from effects caused by changing environmental conditions through the use of protective housings or relatively inert material. Invar steel is one material that is not greatly affected by temperature change. Where protective measures have not been used, environmental effects must be taken into account or the data may not be useful. Additional information on data processing and presentation may be found in EM 1110-2-1908, Rock Testing Handbook, Hanna (1973) and Dunnicliff (1988).

10-17. DATA USE.

An instrumentation program can easily fail if the obtained data is never understood and used. A clear understanding of the purpose of the program is necessary for understanding of the data obtained. Some idea of the behavior that is expected of the structure, usually developed during design and adjusted during construction, is necessary to evaluate the actual behavior. This predicted behavior is the starting point from which all interpretations are made. With these ideas in mind, instrumentation data should prove to be a helpful tool in clearly understanding and evaluating the behavior of any rock foundation or slope.

CHAPTER 11

CONSTRUCTION CONSIDERATIONS

11-1. SCOPE.

This chapter provides general guidance for factors to be considered in the construction of foundations and cut slopes excavated in rock masses. The chapter is divided into five sections with general topic areas to include: Excavation; Dewatering and Ground Water Control; Ground Control; Protection of Sensitive Foundation Materials; and Excavation Mapping and Monitoring.

SECTION I. EXCAVATION

11-2. INFORMATION REQUIREMENTS.

The factors that should be considered when determining the applicability of an excavation method fall into two groups. The first group includes the characteristics of the rock mass to be excavated. The more important of these characteristics are hardness or strength of the intact rock and the degree of fracturing, jointing, bedding, or foliation of the rock mass. This information will normally have been acquired during routine exploration. The second group of factors includes features of the foundation design. These features are the size and shape of the excavation, the tolerances required along the excavation lines, and any restrictions on the time allowed for the excavation to be completed. This second group of factors determines the amount of material to be excavated, the required rate of excavation, the type of finished excavation surface the work must produce, and the amount of working space available.

11-3. EXCAVATION METHODS.

A number of methods are available for excavating rock. These methods include drill and blast, ripping, sawing, water jets, roadheaders, and other mechanical excavation methods.

A. Drill and Blast. Drill and blast is the most common method of excavating large volumes of rock. The hardness of some rock types may eliminate most other excavation techniques from consideration for all but the smallest excavations. Blasting methods can be adapted to many variations in site conditions. Drill and blast techniques, materials, and equipment are thoroughly discussed in EM 1110-2-3800 and the Blasters Handbook (Dupont de Nemours and Company 1977). Due to the availability of that manual, the basics of blasting will not be discussed here. The emphasis of this section will be on aspects of design and construction operations that must be considered when blasting is to be used as a foundation excavation method.

(1) Minimizing foundation damage. Blasting may damage and loosen the final rock surfaces at the perimeter and bottom of the excavation. Although this damage cannot be eliminated completely, in most cases, it can be limited by using controlled blasting techniques. The more common of these techniques are presplitting, smooth blasting, cushion blasting, and line drilling.

a. When presplitting, a line of closely spaced holes is drilled and blasted along the excavation line prior to the main blast. This process creates a fracture plane between the holes that dissipates the energy from the main blast and protects the rock beyond the excavation limits from damage.

b. For the smooth blasting method, the main excavation is completed to within a few feet of the excavation perimeter. A line of perimeter holes is then drilled, loaded with light charges, and fired to remove the remaining rock. This method delivers much less shock and, hence, less damage to the final excavation surface than presplitting or conventional blasting due to the light perimeter loads and the high degree of relief provided by the open face.

c. Cushion blasting is basically the same as smooth blasting. However, the hole diameter is substantially greater than the charge diameter. The annulus is either left empty or filled with stemming. The definitions of smooth and cushion blasting are often unclear and should be clearly stated in any blasting specifications.

d. When using the line drilling method, primary blasting is done to within two to three drill hole rows from the final excavation line. A line of holes is then drilled along the excavation line at a spacing of two to four times their diameter and left unloaded. This creates a plane of weakness to which the main blast can break. This plane also reflects some of the shock from the main blast. The last rows of blast holes for the main blast are drilled at reduced spacing and are lightly loaded. Line drilling is often used to form corners when presplitting is used on the remainder of the excavation.

e. To minimize damage to the final foundation grade, generally blast holes should not extend below grade. When approaching final grade, the rock should be removed in shallow lifts. Charge weight and hole spacing should also be decreased to prevent damage to the final surface. Any final trimming can be done with light charges, jackhammers, rippers, or other equipment. In certain types of materials, such as hard massive rock, it may be necessary to extend blast holes below final grade to obtain sufficient rock breakage to excavate to final grade. This procedure will normally result in overbreak below the final grade. Prior to placing concrete or some types of embankment material, all loose rock fragments and overbreak must be removed to the contractual standard, usually requiring intense hand labor. The overexcavated areas are then backfilled with appropriate materials.

(2) Adverse effects of blasting. Blasting produces ground vibrations, airblast, and flyrock that affect the area around the site. These effects should be kept to a minimum so that nearby structures and personnel are not damaged or injured, and complaints from local residents are kept to a minimum.

a. Ground vibration is the cause of most complaints and structural damage. Ground vibration is usually expressed in terms of peak particle velocity, which can be estimated for a certain location using the equation

$$V = H(D/W^{1/2})^{-B} \qquad (11\text{-}1)$$

where

V = peak particle velocity in one direction, inches per second (ips);

D = distance from blast area to point particle velocity of measurement, ft;

W = charge weight per delay, lb; and

H, B = constants. The constants, H and B, are site-specific and must be determined by conducting test blasts at the site and measuring particle velocities with seismographs at several different distances in different directions. By varying the charge weight for each blast, a log-log plot of peak particle velocity versus scaled distance $(D/W^{1/2})$ may be constructed. The slope of a best fit straight line through the data is equal to the constant B and the value of velocity at a

scaled distance of 1 is equal to the constant H. After determining the constants H and B, Equation 11-1 can then be used to estimate the maximum charge weight that can be detonated without causing damage to nearby structures. If test blasts are not conducted at the site to determine the propagation constants, the maximum charge weight may be estimated by assuming a value for the scaled distance. A value of $50\,\text{ft/lb}^{1/2}$ is considered a minimum safe scaled distance for a site for which no seismograph information is available. Using this value,

$$D/W^{1/2} = 50 \text{ ft/lb}^{1/2} \qquad (11\text{-}2)$$

and

$$W = \left(\frac{D}{50}\right)^2 \qquad (11\text{-}3)$$

where W = maximum safe charge weight per delay in pounds. The maximum safe peak particle velocity for most residential structures is approximately 2 ips. Ground vibration exceeding this level may result in broken windows, cracked walls or foundations, or other types of damage. Blasts fired with a high degree of confinement, such as presplit blasts, may cause higher particle velocities than those predicted by the vibration equation. This is due to the lack of relief normally provided by a free excavation face.

b. Airblast, or compression waves travelling through air, may sometimes damage nearby structures. Noise is that portion of the airblast spectrum having wave frequencies of 20-20,000 Hz. Atmospheric overpressure is caused by the compression wave front. This overpressure may be measured with microphones or piezoelectric pressure gauges. An overpressure of 1 psi will break most windows and may crack plaster. Well-mounted windows are generally safe at overpressures of 0.1 psi, and it is recommended that overpressures at any structure not exceed this level. Airblast is increased by exposed detonating cord, lack of sufficient stemming in blast holes, insufficient burden, heavy low-level cloud cover, high winds, and atmospheric temperature inversions. All of these conditions should be avoided during blasting. Temperature inversions are most common from 1 hr before sunset to 2 hr after sunrise. Blasting should be avoided during these hours if airblast is a concern.

c. Flyrock is usually caused by loading holes near the excavation face with too heavy a charge or by loading explosives too close to the top of the holes. These conditions should be avoided at all times. Flyrock may also be controlled with blast mats. These are large woven mats of wire or rope that are laid over the blast holes or on the face to contain flying debris. Blast mats should be used when blasting very close to existing structures. Extreme caution must be used when placing blast mats to prevent damage to exposed blasting circuits. An alternative to a blasting mat is to place a layer of soil a few feet thick over the blast area prior to blasting to contain the flyrock.

d. Complaints or claims of damage from nearby residents may be reduced by designing blasts to minimize the adverse effects on the surrounding area as much as possible while still maintaining an economic blasting program. To aid in the design of the production blast, test blasts should be conducted and closely monitored to develop attenuation constants for the site. The test blasts should be conducted at several loading factors in an area away from the production blast area or at least away from critical areas of the excavation. However, even with careful blast design, some claims and complaints will most likely occur. People may become alarmed or claim damage when vibration and airblast levels are well below the damage threshold. There are several steps that may be taken to protect against fraudulent or mistaken damage claims. The most basic step is to maintain accurate records of every blast. The blasting contractor is required to submit a detailed blast plan far enough in advance of each shot to allow review by the Government inspector. The blast plan should give all the details of the blast design. After each blast, the contractor should submit a blast report giving the details of the actual blast layout, loading, results, and all other pertinent data. A blast plan and report are normally required on Corps projects. The ground vibrations and airblast from each blast may also be recorded at the nearest structures in several different directions. The seismograph records can be used in the event of a claim to determine if ground vibrations may have reached potentially damaging levels. It is also good practice to record all blasts on videotape. A videotaped record can be helpful in solving various problems with the blasting operations. These monitoring records should be kept, along with the blast plans and records, as a record of the conditions and results of each blast.

e. A further precaution to be taken to protect against damage claims is to require that the contractor perform a preblasting survey of structures near the blasting area. The purpose of the survey is to determine the condition of nearby structures prior to

blasting. The survey should include recording all cracks in plaster, windows, and foundations and photographing the buildings inside and out. The preblast survey might also include basic water quality analyses from any wells in the area. It should also be determined during the survey if there is any sensitive or delicate equipment in nearby buildings that may limit the acceptable peak particle velocity to a value less than the normal 2 ips. This survey should be done at no cost to the property owners. If any property owner refuses to allow his property to be inspected, he should be asked to sign a statement simply stating that he declined the service. The results of the survey will help in determining if damage was preexisting or is blast-related. The scope of the test blasting, monitoring, and preblast survey will be dependent on the size and duration of the production blasting and the anticipated sensitivity of the area as determined by the population density and other social and environmental factors.

f. The key to blasting safety is experienced, safety-conscious personnel. All field personnel directly involved with a blasting operation must be thoroughly familiar with the safety rules and regulations governing the use of explosives. Information and rules on blasting safety are available from explosives manufacturers or the Institute of Makers of Explosives. Safety regulations that apply to Corps of Engineers projects are stated in EM-385-1-1, Safety and Health Requirements, Section 25. These regulations shall be strictly adhered to under all circumstances. The contractor should be required to conduct operations in compliance with all safety regulations. Any unsafe practices must be immediately reported and corrected to avoid accidents.

B. Ripping. Ripping is a means of loosening rock so it may be excavated with loaders, dozers, or scrapers. It involves the use of one or more long narrow teeth that are mounted behind a crawler tractor. Downward pressure is exerted by the tractor and the teeth are pulled through the rock. In addition to standard rippers, impact rippers have been developed in recent years that are capable of breaking relatively strong rock.

(1) Factors influencing rippability. The rock's susceptibility to ripping is related to the rock structure and hardness. The rock structure, in the form of joints, fractures, bedding, faulting, or other discontinuities, determines, to a large degree, the rippability of the rock mass. These discontinuities represent planes of weakness along which the rock may separate. Rock with closely spaced, continuous, near horizontal fractures is much more easily ripped than rock with widely spaced, discontinuous, high-angle fractures. Rock hardness influences the rippability by

determining the amount of force that must be exerted by a ripper tooth to fracture the intact rock. Rock type, fabric, and weathering can be related to the rippability of a rock mass because of the influence they have on the rock structure and hardness. Sedimentary rocks are generally easiest to rip because of their laminated structure. Igneous rock are generally difficult to rip because they are usually hard and lack well-developed lamination. Any weathering that takes place reduces the hardness of the rock and creates additional fractures, making the rock easier to rip. Due to its lesser degree of homogeneity, rock with a coarse-grained fabric is generally weaker than fine grained rock. Because of this, coarse-grained rock is usually easier to excavate by ripping than finer grained rock types.

(2) Rippability indicators. Seismic wave velocity is often used as an indicator of the rippability of a rock mass. The seismic wave velocity is dependent on the rock density or hardness and the degree of fracturing. Hard, intact rock has a higher seismic velocity than softer, fractured rock. Therefore, rocks with lower seismic velocities are generally more easily ripped than those with higher seismic velocities. The seismic wave velocity may be measured using a refraction seismograph and performing a seismic survey of the excavation site. To determine the rippability of the rock, the seismic wave velocity must then be compared with the seismic wave velocities of similar materials in which ripper performance has been demonstrated. Tractor manufacturers have published charts showing, for a particular size tractor and specific ripper configuration, the degree of rippability for different rock types with varying seismic velocities. The rippability of a rock mass may also be assessed by using a rock mass rating system developed by Weaver (1975). Using this system, various rock mass parameters are assigned numerical ratings. The numerical values are then added together to give a rippability rating. Lower ratings indicate easier ripping. Using tractor manufacturer's charts, this rating can be correlated to production rate for various tractor sizes.

(3) Contract considerations. It should never be stated in contract specifications or other legal documents that a rock is rippable or inability to rip designates a new pay item without specifying the tractor size, ripper configurations, and cubic yards loosened per hour (for pay purposes) for which the determination of rippability was made. Rock that may be rippable using a very large tractor may not be rippable using smaller equipment. Not including this qualifying information may lead to claims by a contractor who, after finding he is unable to rip the rock with the size equipment he has available,

claims the contract documents are misleading or incorrect.

(4) Other considerations. Ripping may be used to remove large volumes of rock in areas large enough to permit equipment access. However, ripping produces very poorly sorted muck with many large blocks of rock. Muck from ripping may require further breaking or crushing to make it suitable for use as fill or riprap.

C. Sawing. Sawing is not a common practice, although it is sometimes used as a way of trimming an excavation in soft rock to final grade. Saws may also be used to cut a slot along an excavation line prior to blasting or ripping as an alternative to line drilling. One of the advantages of sawing is that it produces a very smooth excavation face with minimal disturbance to the remaining rock. It also gives very precise control of the position of the final excavation face and may be used to finish fairly complex excavation shapes. Coal saws have been used for sawing soft rocks. Concrete saws may be used for very small-scale work in harder material.

D. Water Jets. High-pressure water jets are beginning to find uses as excavation tools in the construction industry. Water jets cut rock through erosion and by inducing high internal pore pressures that fail the rock in tension. Water jets may range from large water cannon to small hand-held guns. Extremely hard rock may be cut with water jets. However, the pressures required to cut hard rock are extremely high. Optimum pressures for cutting granite may be as high as 50,000 psi. Water jets may be used for cutting slots, drilling holes, trimming to neat excavation lines, and cleaning loose material from an excavated surface. Drill holes may also be slotted or belled. Water jets may not be suitable for use in formations that are extremely sensitive to changes in moisture.

E. Roadheaders. Roadheaders, which are often used in underground excavation, may also be used for final trimming of surface excavations. Roadheaders can rapidly and accurately excavate rock with little disturbance to the remaining rock mass. However, due to power and thrust limitations, their use is limited to rock with an unconfined compressive strength less than approximately 12,000 psi. Large machines may have very high electrical power requirements. Cutting capabilities, length of reach, and power requirements vary widely between models and manufacturers.

F. Other Mechanical Excavation Methods. Various types of mechanical impactors or borehole devices are sometimes used in rock excavation. Mechanical impactors may include hand-operated jackhammers, tractor-mounted rock breakers, or boom-mounted hydraulic impact hammers. These all use chisel or conical points that are driven into the rock by falling weights or by hydraulic or pneumatic hammers. Wedges or hydraulic borehole jacks may be driven or expanded in boreholes to split the rock. Chemicals have been developed that are placed in boreholes much like explosives and, through rapid crystal growth, expand and fracture the rock. Wedges, borehole jacks, or expanding chemicals may provide alternative means of excavation in areas where the vibration and noise associated with blasting cannot be tolerated because of nearby structures or sensitive equipment. Because of their generally low production rates, these alternative methods are normally used only on a limited basis, where excavation quantities are small, or for breaking up large pieces of muck resulting from blasting or ripping. Crane-mounted drop balls are also often used for secondary muck breakage. Jackhammers may be used in confined areas where there is not sufficient room for most equipment to operate.

11-4. EFFECTS OF DISCONTINUITIES ON EXCAVATION.

A. Overbreak. The amount of overbreak, or rock breakage beyond intended excavation lines, is strongly affected by the number, orientation, and character of the discontinuities intersecting the faces of the excavation. Discontinuities represent preferred failure planes within the rock mass. During excavation, the rock will tend to break along these planes. In rock with medium to closely spaced joints that intersect the excavation face, overbreak will most likely occur and will produce a blocky excavation surface. If joints run roughly parallel to the excavation face, overbreak may occur as slabbing or spalling. Worsey (1981) found that if a major joint set intersected the excavation face at an angle less than 15 deg, presplit blasting had little or no beneficial effect on the slope configuration. When blasting, overbreak will also be more severe at the corners of an excavation. Overbreak increases construction costs by increasing muck quantities and backfill or concrete quantities. Because of this, the excavation should be planned and carried out in a way that limits the amount of overbreak. Special measures may be required in areas where overbreak is likely to be more severe because of geologic conditions or excavation geometry. These measures may include controlled blasting techniques or changes in the shape of the excavation.

B. Treatment of Discontinuities. Sometimes, open discontinuities must be treated to strengthen the foundation or prevent underseepage. Open discontinuities encountered in boreholes below the depth of excavation may be pressure-grouted. Open joints and fractures, solution cavities, faults, unbackfilled exploratory holes, or isolated areas of weathered or otherwise unacceptable rock may be encountered during the excavation process. These features must be cleaned out and backfilled. When these features are too small to allow access by heavy equipment normally used for excavation, all work must be done by hand. This process is referred to as dental treatment. Any weathered or broken rock present in the openings is removed with shovels, hand tools, or water jets. The rock on the sides of the opening should be cleaned to provide a good bond with the concrete backfill. Concrete is then placed in the opening, usually by hand.

SECTION II. DEWATERING AND GROUND-WATER CONTROL

11-5. PURPOSES.

Dewatering of excavations in rock is performed to provide dry working conditions for men and equipment and to increase the stability of the excavation or structures. Most excavations that are left open to precipitation or that extend below ground water will require some form of dewatering or ground-water control. Evaluation of the potential need for dewatering should always be included in the design of a structure. Construction contract documents should point out any known potential dewatering problems by the field investigation work.

11-6. PLANNING CONSIDERATIONS.

The complexity of dewatering systems varies widely. Small shallow excavations above ground water may require only ditches to divert surface runoff, or no control at all, if precipitation and surface runoff will not cause significant construction delays. Extensive dewatering systems utilizing several water control methods may be required for larger deeper excavations where inflow rates are higher and the effects of surface and ground-water intrusion are more severe. It must be determined what ground-water conditions must be maintained during the various stages of the construction of the project. The dewatering system must then be designed to establish and maintain those conditions effectively and economically. The size and depth of the excavation, the design and functions of the planned structure, and the project construction and operating schedule must all be considered when evaluating dewatering needs and methods. The dewatering methods must also be compatible with the proposed excavation and ground support systems. The dewatering system should not present obstacles to excavation equipment or interfere with the installation or operation of the ground support systems. The rock mass permeability and existing ground-water conditions must be determined to evaluate the need for, or adequately design, a ground-water control system. The presence and nature of fracture or joint-filling material and the hardness or erodibility of the rock should also be determined to assess the potential for increasing flows during dewatering due to the enlargement of seepage paths by erosion.

11-7. DEWATERING METHODS.

Dewatering refers to the control of both surface runoff and subsurface ground water for the purpose of enhancing construction activities or for improving stability.

A. Surface Water Control. Runoff and other surface waters should be prevented from entering the excavation by properly grading the site. Ditches and dikes may be constructed to intercept runoff and other surface water and direct it away from the work area. Ponding of water on the site should be prevented. Ponded water may infiltrate and act as a recharge source for ground-water seepage into the excavation.

B. Ground-water Control. Ground water may be controlled by a number of different methods. The more commonly used methods include open pumping, horizontal drains, drainage galleries, wells, and cutoffs.

(1) Open pumping. When dewatering is accomplished with the open pumping method, ground water is allowed to enter the excavation. The water is diverted to a convenient sump area where it is collected and pumped out. Collector ditches or berms constructed inside the excavation perimeter divert the water to sumps. Pumps are placed in pits or sumps to pump the water out of the excavation. Most large excavations will require some form of open pumping system to deal with precipitation. In hard rock with clean fractures, fairly large ground-water flows can be handled in this manner. However, in soft rock or in rock containing soft joint filling material, water flowing into the excavation may erode the filling material or rock and gradually in-

crease the size of the seepage paths, allowing flows to increase. Other conditions favorable for the use of open pumping are low hydraulic head, slow recharge, stable excavation slopes, large excavations, and open unrestricted work areas. Open-pumping dewatering systems are simple, easily installed, and relatively inexpensive. However, dewatering by open pumping does not allow the site to be drained prior to excavation. This may result in somewhat wetter working conditions during excavation than would be encountered if the rock mass were pre-drained. Another disadvantage is that the water pressure in low permeability rock masses may not be effectively relieved around the excavation. This method should not be used without supplementary systems if the stability of the excavation is dependent on lowering the piezometric head in the surrounding rock mass. Because the drainage system lies inside the excavation, it may interfere with other construction operations. In some cases, it may be necessary to overexcavate to provide space for the drainage system. If overexcavation is required, the cost of the system may become excessive.

(2) Horizontal drains. Horizontal drains are simply holes drilled into the side of the excavation to intercept high angle fractures within the rock mass. The drain holes are sloped slightly toward the excavation to allow the water to drain from the fractures. The drains empty into ditches and sumps and the water is then pumped from the excavation. This is a very effective and inexpensive way to relieve excess pore pressure in the rock mass behind the excavation sides or behind a permanent structure. The drain holes can be drilled as excavation progresses downward and do not interfere with work or equipment operation after installation. When laying out drain hole locations, the designers must make sure they will not interfere with rockbolts or concrete anchors.

(3) Drainage galleries. Drainage galleries are tunnels excavated within the rock mass outside the main excavation. Drainage galleries normally are oriented parallel to the excavation slope to be drained. Radial drain holes are drilled from the gallery to help collect the water in fractures and carry it into the drainage gallery, where it is then pumped out. Drainage galleries must be large enough to permit access of drilling equipment for drilling the drain holes and future rehabilitation work. This method is effective in removing large quantities of water from the rock mass. Drainage galleries can be constructed prior to the foundation excavation using conventional tunnel construction methods to predrain the rock mass, and they may be utilized as a permanent part of the drainage system for a large project. However, they are very expen-

sive to construct and so are used only when water must be removed from a large area for extended periods of time.

(4) Wells. Pumping wells are often used to dewater excavations in rock. Wells can be placed outside the excavation so they do not interfere with construction operations. Wells also allow the rock mass to be predrained so that all excavation work is carried out under dry conditions. Wells are capable of producing large drawdowns over large areas. They are also effective for dewatering low- to medium-angle fractures that may act as slide planes for excavation slope failures. They will not effectively relieve the pore pressure in rock masses in which the jointing and fracturing is predominantly high angled. The high-angle fractures are not likely to be intersected by the well and so will not be dewatered unless connected to the well by lower angled fractures or permeable zones. The operating cost of a system of pumping wells can be high due to the fact that a pump must operate in each well. Power requirements for a large system can be very high. A backup power source should always be included in the system in the event of failure of the primary power source. Loss of power could result in failure of the entire system.

(5) Cutoffs. Ground-water cutoffs are barriers of low permeability intended to stop or impede the movement of ground water through the rock mass. Cutoffs are usually constructed in the form of walls or curtains.

a. Grouting is the most common method of constructing a cutoff in rock. A grout curtain is formed by pressure-grouting parallel lines of drill holes to seal the fractures in the rock. This creates a solid mass through which ground water cannot flow. However, complete sealing of all fractures is never achieved in grouting. The effectiveness of a grout curtain is difficult to determine until it is in operation. Measurements of changes in grout injection quantities during grouting and pumping tests before and after grouting are normally used to estimate the effectiveness of a grouting operation. Grouting for excavation dewatering can normally be done outside the excavation area and is often used to reduce the amount of water that must be handled by wells or open pumping. It is also used to construct permanent seepage cutoffs in rock foundations of hydraulic structures. Corps of Engineers publications on grouting include EM 1110-2-3506, EM 1110-2-3504, and Albritton, Jackson, and Bangert (1984) (TR GL-84-13).

b. Sheet pile cutoffs may be used in some very soft rocks. However, sheet piling cannot be driven into harder materials. The rock around the sheet pile cutoff may be fractured by the pilings during installation. This will increase the amount of flow around and beneath the cutoff wall and greatly reduce its effectiveness.

c. Slurry walls may also be used as cutoffs in rock. However, due to the difficulty and expense of excavating a deep narrow trench in rock, slurry walls are usually limited to use in soft rocks that may be excavated with machinery also used in soils.

d. Recent developments in mechanical rock excavators that permit excavation of deep slots in relatively strong and hard rock have resulted in increased cost-effectiveness of using diaphragm walls as effective cutoff barriers.

e. Ground freezing may be used to control water flows in areas of brecciated rock, such as fault zones. The use of freezing is generally limited to such soil-like materials. The design, construction, and operation of ground freezing systems should be performed by an engineering firm specializing in this type of work.

SECTION III. GROUND CONTROL

11-8. STABILITY THROUGH EXCAVATION PLANNING.

During the design or construction-planning stages of a project that involve significant cuts in rock, it is necessary to evaluate the stability of the planned excavations. The stability of such excavations is governed by the discontinuities within the rock mass. The occurrence, position, and orientations of the prominent discontinuities at a site should be established during the exploration phase of the project. Using the information and the proposed orientations of the various cut faces to be established, vector analysis or stereonet projections may be used to determine in which parts of the excavation potentially unstable conditions may exist. If serious stability problems are anticipated, it may be possible to change the position or orientation of the structure or excavation slope to increase the stability. However, the position of the structure is usually fixed by other factors. It may not be practical to change either its position or orientation unless the stability problems created by the excavation are so severe that the cost of the necessary stabilizing measures becomes excessive. It may never be possible to delineate all discontinuities and potentially unstable areas before excavation begins. Unexpected problems will likely always be exposed as construction progresses and will have to be dealt with at that time. But performing this relatively simple and inexpensive analysis during design and planning can reduce construction costs. The costs and time delays caused by unexpected stability problems or failures during construction can be extreme. The level of effort involved in determining the stability of the excavation slopes will be governed by the scale of the project and the consequences of a failure. A very detailed stability analysis may be performed for a dam project involving very deep foundation cuts where a large failure would have a serious impact on the economics and safety of the operation. The level of effort for a building with a shallow foundation may include only a surface reconnaissance survey of any exposed rock with minimal subsurface investigations, and then any unstable portions of the excavation may be dealt with during construction.

11-9. SELECTION OF STABILIZATION MEASURES.

When choosing a stabilization method, it is important that the applicable methods be compared based on their effectiveness and cost. In some cases, it may be permissible to accept the risk of failure and install monitoring equipment to give advance warning of an impending failure. Hoek and Bray (1977) gives a practical example of selecting a stabilization method from several possible alternatives.

11-10. STABILIZATION METHODS.

Remedial treatment methods for stabilizing slopes excavated in rock were briefly discussed in Chapter 8. Stabilization methods to include drainage slope configuration, reinforcement, mechanical support, and shotcrete are discussed in more detail below.

A. Drainage. The least expensive method of increasing the stability of a slope is usually to drain the ground water from the fractures. This can be done by horizontal drain holes drilled into the face, vertical pumping wells behind the face, or drainage galleries within the slope. In conjunction with drainage of the ground water, surface water should be kept from entering the fractures in the slope. The ground surface behind the crest should be sloped to prevent pooling and reduce infiltration. Diversion ditches may also be constructed to collect runoff and carry it away from the slope. Diversion and collection ditches should be lined if constructed in highly permeable or moisture sensitive materials.

B. Slope Configuration. Other stabilization methods involve excavating the slope to a more stable configuration. This can be done by reducing the slope angle or by benching the slope. Benching results in a reduced overall slope angle, and the benches also help to protect the work area at the base of the slope from rockfall debris. If the majority of the slope is stable and only isolated blocks are known to be in danger of failing, those blocks may simply be removed to eliminate the problem. The use of controlled blasting techniques may also improve the stability of an excavated slope by providing a smoother slope face and reducing the amount of blast-induced fracturing behind the face.

C. Rock Reinforcement. Rock reinforcement may be used to stabilize an excavation without changing the slope configuration and requiring excess excavation or backfill. Rock bolts or untensioned dowels are used to control near surface movements and to support small- to medium-size blocks. They may be installed at random locations as they are needed or in a regular pattern where more extensive support is required. Rock anchors or tendons are usually used to control movements of larger rock masses, because of their greater length and higher load capacity. One of the advantages of using reinforcement is that the excavation face may be progressively supported as the excavation is deepened. Thus, the height of slope that is left unsupported at any one time is equal to the depth of a single excavation lift or bench. After installation, rock reinforcement is also out of the way of activity in the work area and becomes a permanent part of the foundation. Rock bolts or anchors may also be installed vertically behind the excavation face prior to excavation to prevent sliding along planar discontinuities that will be exposed when the cut face is created. The effects of rock reinforcement are usually determined using limit equilibrium methods of slope analysis. Methods for determining anchorage force and depth are given in Chapter 9 on anchorage systems. While the methods discussed in Chapter 9 were primarily developed for calculating anchor forces applicable to gravity structures, the principles involved are also applicable to rock slopes. Additional information may be found in EM 1110-1-2907 (1980) and in the references cited in Chapter 8.

D. Mechanical Support and Protection Methods. Mechanical support methods stabilize a rock mass by using structural members to carry the load of the unstable rock. These methods do not strengthen the rock mass. The most common type of mechanical support for foundation excavations is bracing or shoring. In rock excavations, support usually consists of steel beams placed vertically against the excavation face. In narrow excavations, such as trenches, the vertical soldier beams are held in place by horizontal struts spanning the width of the trench. In wider excavations, the soldier beams are supported by inclined struts anchored at the lower end to the floor of the excavation. Steel or timber lagging may be placed between the soldier beams where additional support is needed. One of the disadvantages of bracing and shoring is that mobility in the working area inside the excavation is hampered by the braces. A common solution to this problem is to tie the soldier beams to the rock face with tensioned rock bolts. This method utilizes the benefits of rock reinforcement while the beams spread the influence of each bolt over a large area. When only small rock falls are expected to occur, it may not be necessary to stabilize the rock. It may be necessary only to protect the work area in the excavation from the falling debris. Wire mesh pinned to the face with short dowels will prevent loose rock from falling into the excavation. The mesh may be anchored only at its upper edge. In this case, the falling debris rolls downslope beneath the mesh and falls out at the bottom of the slope. Wire mesh may be used in conjunction with rock bolts and anchors or bracing to help protect workers from debris falling between larger supports. Buttresses, gabions, and retaining walls, although commonly used for support of permanent slopes, are not normally used to support temporary foundation excavations.

E. Shotcrete. The application of shotcrete is a very common method of preventing rock falls on cut rock slopes. Shotcrete improves the interlock between blocks on the exposed rock surface. The shotcrete does not carry any load from the rock and so is more a method of reinforcement than of support. Shotcrete may also be applied over wire mesh or with fibers included for added strength and support. Shotcrete is fast and relatively easy to apply and does not interfere with workings near the rock cut. Shotcrete also aids in stabilizing rock cuts by inhibiting weathering and subsequent degradation of the rock. This is discussed further in Section IV on protection of sensitive foundations.

SECTION IV. PROTECTION OF SENSITIVE FOUNDATION MATERIALS

11-11. GENERAL.

Some rocks may weather or deteriorate very rapidly when exposed to surface conditions by excavation processes. These processes may cause a considerable decrease in the strength of the near surface materials. The processes most likely to be responsible for such damage are freeze/thaw, moisture loss or gain, or chemical alteration of mineral constituents. To preserve the strength and character of the foundation materials, they must be protected from damaging influences.

11-12. COMMON MATERIALS REQUIRING PROTECTION.

There are several rock types that, because of their mineralogy or physical structure, must be protected to preserve their integrity as foundation materials.

A. Argillaceous Rocks. Shales and other argillaceous rocks may tend to slake very rapidly when their moisture content decreases because of exposure to air. This slaking causes cracking and spalling of the surface, exposing deeper rock to the drying effects of the air. In severe cases, an upper layer of rock may be reduced to a brecciated, soil-like mass.

B. Swelling Clays. Joint filling materials of montmorillonitic clays will tend to swell if their moisture content is increased. Swelling of these clays brought about by precipitation and runoff entering the joints may cause spalling or block movement perpendicular to the joints.

C. Chemically Susceptible Rock. Some rock types contain minerals that may chemically weather at a very rapid rate to a more stable mineral form. The feldspars in some igneous rocks and the chlorite and micas in some schists may rapidly weather to clays when exposed to air and water. This process can produce a layer of clayey, ravelling material over the surface of hard, competent rock.

D. Freeze/Thaw. Most rocks are susceptible to some degree to damage from freezing. Water freezing in the pores and fractures of the rock mass may create high stresses if space is not available to accommodate the expansion of the ice. These high stresses may create new fractures or enlarge or propagate existing fractures, resulting in spalling from the exposed face.

11-13. DETERMINATION OF PROTECTION REQUIREMENTS.

The susceptibility of the foundation materials to rapid deterioration or frost damage should be determined during the exploration phase of a project. If possible, exposures of the materials should be examined and their condition and the length of time they have been exposed should be noted. If core samples are taken as part of the exploration program, their behavior as they are exposed to surface conditions is a very good indication of the sensitivity of the foundation materials to moisture loss. Samples may also be subjected to freeze/thaw and wet/dry cycles in the laboratory. The behavior of the rock at projects previously constructed in the same materials is often the best source of information available, provided the construction process and schedule are similar. In this respect, the project design, construction plan, and construction schedule play important roles in determining the need for foundation protection. These determine the length of time excavated surfaces will be exposed. Climatic conditions during the exposure period will help determine the danger of damage from frost or precipitation.

11-14. FOUNDATION PROTECTION METHODS.

The first step in preventing damage to sensitive foundation materials is to plan the construction to minimize the length of time the material is exposed. Construction specifications may specify a maximum length of time a surface may be exposed without requiring a protective coating. Excavation may be stopped before reaching final grade or neat excavation lines if a surface must be left exposed for an extended period of time. This precaution is particularly wise if the material is to be left exposed over winter. The upper material that is damaged by frost or weathering is then removed when excavation is continued to final profiles and the rock can be covered more quickly with structural concrete. It may not be possible to quickly cover the foundation materials with structural concrete. In this case, it is necessary to temporarily protect the foundation from deterioration. This can be done by placing a protective coating over the exposed foundation materials.

A. Shotcrete. Sprayed-on concrete, or shotcrete, is becoming perhaps the most common protective coating for sensitive foundation materials. Its popularity is due largely to the familiarity of engineers, inspectors, and construction contractors with its design and application. Shotcrete can be easily and quickly applied to almost any shape or slope surface. If correctly applied, it prevents contact of the rock with air and surface water. If ground water is seeping from the rock, weep holes should be made in the shotcrete to help prevent pressure buildup between the rock and the protective layer. Otherwise, spalling of the shotcrete will most likely occur. The shotcrete may be applied over wire mesh

pinned to the rock to improve the strength of the protective layer. When used as a protective coating only, the thickness of the shotcrete will normally be 2-3 in.

B. Lean Concrete or Slush Grouting. Slush grouting is a general term used to describe the surface application of grout to seal and protect rock surface. The grout used is usually a thin sand cement grout. The mix is spread over the surface with brooms, shovels, and other hand tools and worked into cracks. No forms of any kind are used. Lean concrete may also be specified as a protective cover. It is similar to slush grouting in that it is placed and spread largely by hand. However, the mix has a thicker consistency and a thicker layer is usually applied. Because of the thicker application, some forming may be necessary to prevent lateral spreading. Both methods provide protection against surface water and moisture loss to the air. The use of slush grouting and lean concrete for protection are limited to horizontal surfaces and slopes of less than about 45 deg due to the thin mixes and lack of forming.

C. Plastic Sheeting. Sheets of plastic, such as polyethylene, may be spread over foundation surfaces to prevent seepage of surface moisture into the rock. This may also provide a small degree of protection from moisture loss for a short time. Sheet plastics work best on low- to medium-angled slopes. The plastic sheets are difficult to secure to steep slopes, and water may stand on horizontal surfaces and penetrate between sheets. The sheets can be conveniently weighted in place with wire mesh.

D. Bituminous Coatings. Bituminous or asphaltic sprays may also be used as protective coatings. These sprays commonly consist of asphalt thinned with petroleum distillates. The mixture is heated to reduce its viscosity and is then sprayed onto the rock surface. These coatings are effective as temporary moisture barriers. However, they are not very durable and usually will not remain effective for more than 2 to 3 days.

E. Resin Coatings. Various synthetic resins are manufactured for use as protective coatings for rock, concrete, and building stone. These products generally form a low permeability membrane when sprayed on a surface. The membrane protects the rock from air and surface water. Life expectancy, mixes, and materials vary with different manufacturers. These materials require specialized equipment and experienced personnel for application. Resin coatings may need to be removed from rock surfaces prior to placement of structural concrete to assure proper rock/concrete bond. Sources of additional information are limited due to the somewhat limited use of these coatings. Potential suppliers of these materials may include manufacturers of coatings, sealers, or resin grouts.

SECTION V. EXCAVATION MAPPING AND MONITORING

11-15. MAPPING.

Geologic mapping should be an integral part of the construction inspection of a foundation excavation. This mapping should be performed by the project geologist who will prepare the Construction Foundation Report required by ER 1110-2-1801. Thorough construction mapping ensures that the final excavation surfaces are examined and so aids in the discovery of any unanticipated adverse geologic conditions. Mapping also provides a permanent record of the geologic conditions encountered during construction. Appendix B of EM 1110-1-1804, Geotechnical Investigations, and Chapter 3 of this report outline procedures for mapping open excavations.

11-16. PHOTOGRAPHY.

Photographs should be taken of all excavated surfaces and construction operations. As with mapping, photographs should be taken by the person(s) responsible for preparation of the Construc-

tion Foundation Report (ER 1110-2-1801). However, project staffing may be limited such that it may be necessary to require the contractor to take the photographs. All photos must be properly labeled with date, subject, direction of view, vantage point, photographer, and any other pertinent information. Photographs of excavated surfaces should be as unobstructed as possible. Complete photographic coverage of the project is very important. Recently, videotaping has also provided benefits. This should be impressed upon the geologists and engineers responsible for construction mapping and inspection.

11-17. CONSTRUCTION MONITORING.

Monitoring of construction procedures and progress should be performed on a regular basis by the designers in accordance with ER 1110-2-112. The schedule of design visits should be included in the Engineering Considerations and Instructions to Field Personnel. Excavation monitoring must be performed as thoroughly and frequently as possible to ensure that complete information is obtained on the as-built condition of the rock foundation. A checklist may be used that allows the inspector to give a brief description of various features of the foundation and the construction activities. An example of such a checklist is given in Appendix B of EM 1110-1-1804.

CHAPTER 12

SPECIAL TOPICS

12-1. SCOPE.

This chapter provides general guidance in recognizing and treating special conditions which can be encountered in rock foundations that cause construction or operation problems. These conditions are likely to be encountered only within certain regions and within certain rock types, but geotechnical professionals should be aware of the potential problems and methods of treatment. This chapter is divided into three topic areas: karst, pseudokarst, and mines that produce substantial underground cavities; swelling and squeezing rock, much of which may be described as a rock but treated as a soil; and gradational soil-rock contacts, rock weathering, saprolites, and residual soils that make determination, selection, and excavation of suitable bearing elevations difficult.

SECTION I. KARST, PSEUDOKARST, AND MINES

12-2. CAVITIES IN ROCK.

A topic of concern in many projects involving rock excavation is whether or not there are undetected cavities below an apparently solid bedrock surface or whether cavities could develop after construction. These cavities may occur naturally in karst or pseudokarst terrains, may be induced by human interference in natural processes, or may be totally due to man's activities. The term "cavities" is used since it covers all sizes and origins of underground openings of interest in rock excavations.

A. Cavity Significance. The presence of cavities has a number of rock engineering implications, including:

- Irregular or potentially irregular bedrock topography due to collapse or subsidence and associated unpredictable bearing surface elevations.
- Excavation difficulties, with extensive hand-cleaning, grouting, and dental treatment requirements.
- Questionable support capacity with a potential for collapse or subsidence over cavities, or settle-

ment of debris piles from prior collapses, all of which may be concealed by an apparently sound bedrock surface.
- Ground water flow problems, with requirements for tracing flow paths, or sealing off or diverting flows around or through the project area. Surface water flows may be affected by underground cavities, sometimes by complete diversion to the subsurface.
- Contaminants may flow rapidly into open channels, with minimal natural filtration and purification, possibly contaminating local water supplies.

B. Problem Rocks.

(1) Most natural and induced cavities develop in soluble rocks, most notably limestone, dolomite, gypsum, and rock salt. Typical karst conditions develop in limestones and dolomites by solution-widening of joints and bedding planes caused by flowing ground water. Eventually, this process develops into a heterogeneous arrangement of cavities with irregular sinkholes occurring where cavity roofs have collapsed. The amount of solution that occurs in limestone and dolomite would be negligible in the lifetime of a typical project. Hence, existing cavities are the major concern.

(2) Gypsum and anhydrite are less common than limestones, but they have the additional concern of solution and collapse or settlement during the useful life of a typical structure. Flow of ground water, particularly to water supply wells, has been known to dissolve gypsum and cause collapse of structures. Rock salt is probably one of the most soluble of common geologic materials, and may be of concern in some areas, particularly along the Gulf of Mexico, the Michigan Basin, and in central Kansas. While natural occurrences of cavities in rock salt are rare, cavities may have been formed by solution mining methods, and collapse or creep has occurred in some of the mined areas.

(3) Pseudokarst terrain is an infrequently encountered form that appears to be classic karst topography, but occurs in a different geologic environment. Cavities and sinkholes can occasionally occur in lava flow tubes or in poorly cemented sandstones adjacent to river valleys or coastlines. The same basic engineering problems and solutions apply to

pseudokarst as to karst topography, but generally on a less severe scale. Care should be taken to avoid attributing surface features to pseudokarst conditions, when true karst conditions in lower rock strata may be the actual cause.

C. Mining Activities. Mining is the principle cause of human-induced cavities, and subsidence or collapse over old mines is one of the oldest forms of surface disruption caused by man. Coal, with occurrences shown in Figure 12-1, is probably the most common material extracted by underground mining, although nearly any valuable mineral may have been mined using any scale of mining operation. The mines typically follow beds or ore bodies that are relatively easy to follow using stratigraphic or structural studies. The actual locations of mined cavities may be more difficult to determine. Mines in recent times generally have excellent layout maps available, but older mines may not be well documented. In some cases, small-scale prospect operations may be totally obscured until excavations are at an advanced stage.

12-3. INVESTIGATIONS.

Cavities are difficult to detect and are undiscovered until exposed by construction excavations. A combination of detailed preconstruction investigations and construction investigations should be anticipated in potential cavity areas. In this respect, karst topography develops in relatively predictable regions of limestones and dolomites, as shown in Figure 12-2 and Table 12-1. However, the occurrence of cavities on a local scale is more difficult to determine, and many significant cavities can be missed by a typical exploration program. The inability to detect specific cavities also holds true for pseudokarst terrains, rock salt, gypsum, and mine cavities. The Geotechnical Investigations Manual, EM 1110-1-1804, provides guidance on the screening of an area for sinkholes, anhydrites or gypsum layers, caves, and area subsidence.

A. Initial Site Investigations. Geophysics may be of some use in initial site investigations in locating larger cavities, but may miss

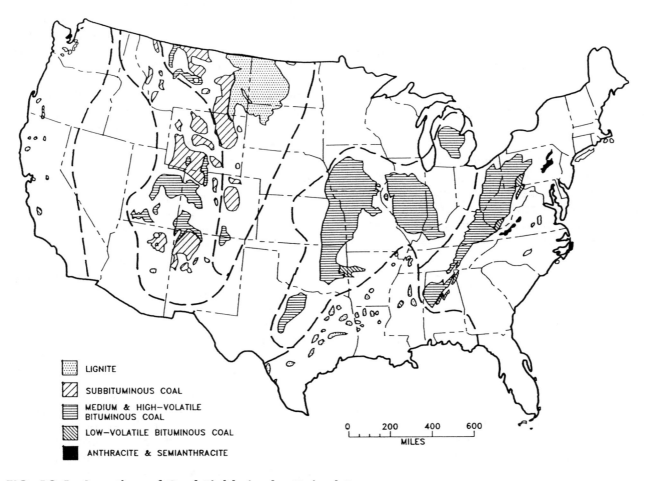

LIGNITE

SUBBITUMINOUS COAL

MEDIUM & HIGH-VOLATILE BITUMINOUS COAL

LOW-VOLATILE BITUMINOUS COAL

ANTHRACITE & SEMIANTHRACITE

0   200   400   600
MILES

FIG. 12-1. Location of Coal Fields in the United States

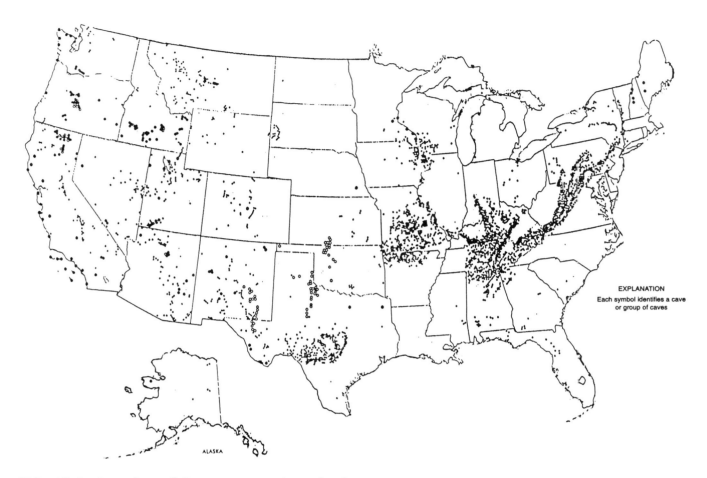

FIG. 12-2. Location of Cavern Areas in United States

smaller ones. Remote sensing using air photos, infrared imagery, and side-looking radar is useful in determining trends of cavities and jointing in an area, as well as determining structural geology features associated with rock salt exposures. Detailed joint strike and dip mapping, in some cases by removing site overburden, may be very useful in predicting the trends of known cavities that follow joints. In some cases, hydrologic testing using piezometers, dye flow tracers, and pump tests may help determine permeabilities and probable flow paths along cavities. In the case of mines, stratigraphic analysis of economic minerals and ore body studies, along with studies of mining company records and Government documents associated with the mine, can help in determining the mine layout. Surveys from inside mines are desirable, but may not be possible due to dangerous conditions. Borehole cameras may be used to determine the size and condition of otherwise inaccessible mines. Table 12-2 shows several exploration and investigation methods that may be of more value in detection of cavities.

B. Cavity Detection. Since cavity occurrence is difficult to determine on a local scale, the only practical solution, after initial site studies, is to place a test boring at the location of each significant load-bearing member. Such an undertaking is costly, but represents the only reasonable approach in areas of high concern.

12-4. ALTERNATIVE SOLUTIONS.

A number of techniques/methods are available for addressing design and construction problems associated with project sites where cavities are present. The following provides a brief listing of alternative techniques.

• Avoid the area for load-bearing use if possible.
• Bridge the cavity by transferring the loads to the cavity sides.
• Allow for subsidence and potentially severe differential settlements in the design of the foundation and structure.
• Fill in the cavities to minimize subsidence, prevent

TABLE 12-1. Summary of Major Karst Areas of United States

Karst area (1)	Location (2)	Characteristics (3)
Southeastern coastal plain	South Carolina, Georgia	Rolling, dissected plain, shallow dolines, few caves; Tertiary limestone generally covered by thin deposits of sand and silt.
Florida	Florida, southern Georgia	Level to rolling plain: Tertiary, flat-lying limestone; numerous dolines, commonly with ponds; large springs; moderate-sized caves, many water-filled.
Appalachian	New York, Vermont, south to northern Alabama	Valleys, ridges, and plateau fronts formed south of Palaeozoic limestones, strongly folded in eastern part; numerous large caves, dolines, karst valleys, and deep shafts; extensive areas of karren.
Highland Rim	Central Kentucky, Tennessee, northern Georgia	Highly dissected plateau with Carboniferous, flat-lying limestone; numerous large caves, karren, large dolines and uvala.
Lexington-Nashville	North-central Kentucky, central Tennessee, southeastern Indiana	Rolling plain, gently arched; Lower Palaeozoic limestone; a few caves, numerous rounded shallow dolines.
Mammoth Cave-Pennyroyal Plain	West-central, southwestern Kentucky, southern Indiana	Rolling plain and low plateau; flat-lying Carboniferous rocks; numerous dolines, uvala and collapse sinks; very large caves, karren developed locally, complex subterranean drainage, numerous large "disappearing" streams.
Ozarks	Southern Missouri, northern Arkansas	Dissected low plateau and plain; broadly arched Lower Palaeozoic limestones and dolomites; numerous moderate-sized caves, dolines, very large springs; similar but less extensive karst in Wisconsin, Iowa, and northern Illinois.
Canadian River	Western Oklahoma, northern Texas	Dissected plain, small caves and dolines in Carboniferous gypsum.
Pecos Valley	Western Texas, southeastern New Mexico	Moderately dissected low plateau and plains; flat-lying to tilted Upper Palaeozoic limestones with large caves, dolines, and fissures; sparse vegetation; some gypsum karst with dolines.
Edwards Plateau	Southwestern Texas	High plateau, flat-lying Cretaceous limestone; deep shafts, moderate-sized caves, dolines; sparse vegetation.
Black Hills	Western South Dakota	Highly dissected ridges; folded (domed) Palaeozoic limestone; moderate-sized caves, some karren and dolines.
Kaibab	Northern Arizona	Partially dissected plateau, flat-lying Carboniferous limestones; shallow dolines, some with ponds; few moderate-sized caves.
Western mountains	Wyoming, northwestern Utah, Nevada, western Montana, Idaho, Washington, Oregon, California	Isolated small areas, primarily on tops and flanks of ridges, and some area in valleys; primarily in folded and tilted Palaeozoic and Mesozoic limestone; large caves, some with great vertical extent, in Wyoming, Utah, Montana, and Nevada; small to moderate-size caves elsewhere; dolines and shafts present; karren developed locally.

TABLE 12-2. Effectiveness of Cavity Investigation Techniques

Cavity type (1)	Increased borings (2)	Investigation method considered[a]						
		Geophysics (3)	Remote sensing (4)	Piezometers (5)	Pump tests dye flow tests (6)	Discontinuity analysis (7)	Borehole cameras (8)	Mine record studies (9)
Anhydrite gypsum	2	1	2	1	1	3	1	2
Karst	5	4	4	2	3	5	5	1
Salt	2	3	3	1	1	1	2	3
Mines	4	4	4	1	1	1	5	5
Lava tubes	4	2	4	2	2	1	3	1

[a]Ratings: Grade from 1 = not effective to 5 = highly effective

catastrophic collapse, and prevent progressive enlargement. Support piers or walls may be used for point supports in larger cavities, or cavities may be filled with sand, gravel, and grout. Cement grout can be used to fill large cavities to prevent roof slabs from falling, eliminating a potential progression to sinkholes. Grout also can fill cavities too small for convenient access, thereby reducing permeability and strengthening the rock foundation.

- Avoid placing structures over gypsum, salt, or anhydrite beds where seeping or flowing water can rapidly remove the supporting rock.
- Plan for manual cleaning of pinnacled rock surfaces with slush grouting and dental treatment of enlarged joints as shown in Figure 12-3. The exact extent of this work is difficult to predict prior to excavation.
- Control surface and ground-water flow cautiously. Lowering of the water table has induced collapses and the formation of new sinkholes in previously unexpected areas. Surface drainage in most karst areas is poorly developed, since most drainage has been to the subsurface.

SECTION II. SWELLING AND SQUEEZING ROCK

12-5. GENERAL.

The case of swelling or squeezing rock represents yet another special problem. In such cases, the rock foundation changes after it is exposed or unloaded, and the rock expands (increases in apparent volume) horizontally or vertically. There are at least five mechanisms that can cause swelling rock.

Swelling may be the result of a single mechanism or a combination of several interacting mechanisms. The five common mechanisms of swelling rock include elasto-plastic rebound (or heave), cation hydration, chemical reaction, loss of internal strength (creep), and frost action. Some of these mechanisms occur most commonly in certain rock types. Each category is discussed individually.

12-6. REBOUND.

Elasto-plastic rebound is the expansion of rock due to the reduction or removal of external forces acting upon the rock mass. In some cases, especially in areas with a high horizontal stress field, removal of as little as a few feet of rock or soil may result in an expansion of the exposed rock. The expansion may be expressed as a general heave of the exposed rock or as a pop-up or buckling. This behavior frequently occurs in areas associated with glacial activity and can occur in most types of rock. In structural excavations where rebound may be a problem, the surface may be rapidly loaded with a weight equivalent to the overburden to prevent rebound of the rock. In many nonstructural open cut excavations, this type of swelling may be more of a minor maintenance problem than a serious concern.

12-7. CATION HYDRATION.

Cation hydration is another mechanism for swelling that is most frequently associated with some argillaceous rocks. The process refers to the attraction and adsorption of water molecules by clay minerals. Factors that contribute to this form of swell include poor cementation, desiccation and rewet-

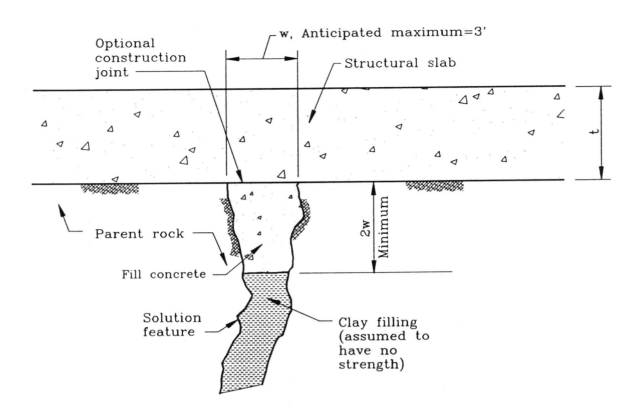

NOTES:
1. CLEAN ALL SOLUTION FEATURES TO A MINIMUM DEPTH OF 2w. IF FEATURE WIDENS WITH DEPTH, CLEAN TO A DEPTH WHERE WEDGING CAN BE ACHIEVED. FINAL DEPTH OF CLEANING FOR ALL FEATURES TO BE COORDINATED WITH PROJECT GEOLOGIST.
2. DEGREE OF CLEANING TO BE THE SAME AS FOR HORIZONTAL FOUNDATION SURFACES.
3. IN THE EVENT THAT A ZONE OF NUMEROUS CLOSELY SPACED SOLUTION FEATURES(NO INDICATION OF THIS FROM DRILLING), w SHOULD BE TAKEN AS THE ZONE WIDTH.
4. FEATURES LESS THAN 3" WIDE CAN BE DISREGARDED EXCEPT FOR NORMAL FOUNDATION CLEANUP.

FIG. 12-3. Criteria for Treatment of Solution-Widened Joints

ting, unloading, and high clay mineral content, especially montmorillonite clay.

A. Problem Rocks. Clay shale is the rock most commonly associated with swelling problems, and its principal mechanism of swell is cation hydration. Defined as shales that tend to slake easily with alternate wetting and drying, clay shales were overconsolidated by high loads in the past. Typically, clay shales were deposited in shallow marine or deltaic environments in Cretaceous or Paleocene times and contain a high percentage of swell prone montmorillonite mineral.

B. Other Factors. Factors other than clay mineral type and content may also contribute to cation hydration induced swell. These factors include density, moisture content, rock mineral structure, loading history, and weathering.

(1) Density of the rock is an important indicator of swell. A 25% increase in the dry density of clay shales can more than double the maximum swell pressures developed in the material. Therefore,

a high density could indicate high swell pressure potential.

(2) Low moisture content can indicate a high swell potential, since there is more availability for water within the clay structure.

(3) The mineral structure of the clay shale can influence the magnitude and isotropy of the swell characteristics. A compacted mineral orientation typical of clay shales has most of the plate mineral faces in a "stacked" arrangement, with maximum swell potential normal to the mineral faces.

(4) The loading history can indicate the degree of preconsolidation that the shales have been subjected to in the past. Changes in the stress environment can be due to erosion, glaciation, stream downcutting, and engineering activities.

(5) Weathering of clay shales generally reduces the swell potential unless additional expansive clays are formed.

C. Excavation Problems. Excavations in clay shale present special problems. If the excavated surface is allowed to dry, the material develops shrinkage cracks, and rebound-type swell induces a relative moisture reduction and density decrease. Water or moisture from concrete applied to this surface can induce swelling of the clay shale. Slope stability is another prominent problem in clay shales, since any excavation can result in renewed movement along older, previously stable slide planes. The presence of unfavorably oriented bentonite seams common in clay shales can present serious stability hazards.

D. Treatment Methods. Preventive measures can include careful control of the excavation sequence, moisture control, and surface protection, and favorable stratigraphic placement and orientation of the slopes and structures. Treatment methods are discussed in Chapter 11.

E. Field Investigations. Field investigations should include checks for significant problem-prone clay shale formations, some of which are listed in Table 12-3. Also, some indications of clay shale swell problems include hummocky terrain along river valley slopes, slides along road cuts, and tilting or cracking of concrete slabs or light structures. Slickensides in shale is another indicator of swelling potential. The presence of clay shales susceptible to cation hydration swelling in the project region should be determined very early in the exploration program.

F. Laboratory Tests. Laboratory tests to determine engineering properties are similar to those for soil mechanics, and include clay mineral type and percentage analyses, Atterberg limits, moisture content, consolidation tests, and swell tests. In decreasing order, the significant swell-producing clay minerals are montmorillonite, illite, attapulgite, and kaolinite. Atterberg limit tests can indicate the swelling nature of clay shale, with high plasticity indices correlating to high swell potential, as shown in Table 12-4. The method of determining the Atterberg limits of clay shale must be consistent, since air drying, blending, or slaking of the original samples may provide variable results. There are several methods of performing consolidation and swell tests on clay shales. These methods are summarized in Table 12-5.

12-8. CHEMICAL-REACTION SWELLING.

Chemical-reaction swelling refers to a mechanism most commonly associated with Paleozoic black shales, such as the Conemaugh Formation, or the Monongahela Formation in Pennsylvania. Swell develops when reactions such as hydration, oxidation, or carbonation of certain constituent minerals create by-products that result in volumes significantly larger than the original minerals. These reactions can result in large swelling deformations and pressures after excavation and construction. The conditions that are conducive to this type of swelling may not occur until after a foundation is in place, and similar conditions may not be reproduced easily in the laboratory to indicate that it may be a problem. Temperature, pressure, moisture, adequate reactants and, in some cases, bacterial action are critical parameters for reaction to occur.

A. Reactions. The transformation of anhydrite to gypsum is one of the more common reactions. In shales containing a substantial percentage of free pyrite, a similar reaction can occur. The oxidation of the pyrite can result in the growth of gypsum crystals or a related mineral, jarosite. The presence of sulphur bacteria can aid the reaction, and may be essential to the reaction in some cases. Sulfuric acid produced by the reaction may react with any calcite in the shale to increase the development of gypsum. The resulting growth of gypsum crystals causes the swelling, which can uplift concrete structures.

B. Treatment Methods. Since the reactions are difficult to predict or simulate during exploration and design, it may be desirable to avoid placing structures on pyrite-bearing carbonaceous shales of Paleozoic age, since these rocks are the most common hosts for chemical-reaction swelling. If avoidance is not an option, the exposed surfaces may be protected from moisture changes by placing a sealing membrane of asphalt or some other suitable material. Shotcrete is not a suitable coating material since the sulfuric acid produced by the reaction can destroy it. The added source of calcium may even enhance the swelling reaction. Another preventive measure may be the application of a chemical additive that blocks the growth of gypsum crystals. Tests have indicated that diethylenetriamine penta (methylene phosphonic acid) substantially inhibited gypsum development under normally reactive conditions. Other similar crystal growth inhibitors may be useful in preventing chemical-reaction swelling.

12-9. LOSS OF INTERNAL STRENGTH.

This swelling mechanism occurs most commonly when intact rock loses its internal bonding or cementation. The mechanism is commonly associ-

TABLE 12-3. Landslide-Susceptible Clay Shales in United States

Stratigraphic unit (1)	Description (2)
Bearpaw shale	Upper Cretaceous; northern, eastern, and southern Montana, central northern Wyoming, and southern Alberta, Canada; marine clay shale 600 to 700 ft; in Montana group.
Carlile shale	Upper Cretaceous; eastern Colorado and Wyoming, Nebraska, Kansas, and South Dakota, southeastern Montana and northeastern New Mexico; shale, 175 to 200 ft; in Colorado group.
Cherokee shale	Early Pennsylvanian; eastern Kansas, southeastern Nebraska, northwestern Missouri, and northeastern Oklahoma; shale, 500 ft; in Des Moines group.
Claggett formation	Upper Cretaceous; central and eastern Montana, and central northern Wyoming; marine clay shales and sandstone beds, 400 ft; in Montana group.
Dawson formation	Upper Cretaceous-Lower Tertiary; central Colorado; nonmarine clay shales, siltstone and sandstone, 1,000 ft.
Del Rio clay	Lower Cretaceous; southern Texas; laminated clay with beds of limestone; in Washita group.
Eden group	Upper Ordovician; southwestern Ohio, southern Indiana, and central northern Kentucky; shale with limestone, 250 ft; in Cincinnati group.
Fort Union group	Paleocene; Montana, Wyoming, North Dakota, northwestern South Dakota, and northwestern Colorado; massive sandstone and shale, 4,000 ft+.
Frontier formation	Upper Cretaceous; western Wyoming and southern Montana; sandstone with beds of clay and shale, 2,000 to 2,600 ft; in Colorado group.
Fruitland formation	Upper Cretaceous; southwestern Colorado and northwestern New Mexico; brackish and freshwater shales and sandstones, 194 to 530 ft; late Montana age.
Graneros shale	Upper Cretaceous; eastern Colorado and Wyoming, southeastern Montana, South Dakota, Nebraska, Kansas, and northeastern New Mexico; argillaceous or clayey shale, 200 to 210 ft; in Colorado group.
Gros Ventre formation	Middle Cambrian; northwestern Wyoming and central southern Montana; calcareous shale with conglomeratic and oolitic limestone, 800 ft.
Jackson group	Upper Eocene; Gulf Coastal Plain (southwestern Alabama to southern Texas); calcareous clay with sand, limestone, and marl beds.
Mancos shale	Upper Cretaceous; western Colorado, northwestern New Mexico, eastern Utah, southern and central Wyoming; marine, carbonaceous clay shale with sand, 1,200 to 2,000 ft; of Montana and Colorado age.
Merchantville clay	Upper Cretaceous; New Jersey; marly clay, 35 to 60 ft; in Matawan group.
Modelo formation	Upper Miocene; southern California; clay, diatomaceous shale, sandstone, and cherty beds, 9,000 ft.
Monterey shale	Upper, middle, and late lower Miocene; western California; hard silica-cemented shale and soft shale, 1,000 ft+.
Morrison formation	Upper Jurassic; Colorado and Wyoming, south central Montana, western South Dakota, western Kansas, western Oklahoma, northern New Mexico, northeastern Arizona, and eastern Utah; marl with sandstone and limestone beds, 200 ft+.
Mowry shale	Upper Cretaceous; Wyoming, Montana, and western South Dakota; hard shale, 150 ft; in Colorado group.
Pepper formation	Upper Cretaceous; eastern Texas; clay shale.
Pierre shale	Upper Cretaceous; North Dakota, South Dakota, Nebraska, western Minnesota, eastern Montana, eastern Wyoming, and eastern Colorado; marine clay shale and sandy shale, 700 ft; in Montana group.
Rincon shale	Middle or lower Miocene; southern California; clay shale with lime stone, 300 to 2,000 ft.
Sundance formation	Upper Jurassic; southwestern South Dakota, Wyoming, central southern Montana; northwestern Nebraska; and central northern Colorado; shale with sandstone, 60 to 400 ft.
Taylor marl	Upper Cretaceous; central and eastern Texas; chalky clay, 1,200 ft.
Thermopolis shale	Upper Cretaceous; central northern Wyoming, and central southern Montana; shale with persistent sandy bed near middle, 400 to 800 ft; in Colorado group.

TABLE 12-3. Landslide-Susceptible Clay Shales in United States (continued)

Stratigraphic unit (1)	Description (2)
Trinity group	Lower Cretaceous; Texas, south central and southeastern Oklahoma, southwestern Arkansas, and northwestern Louisiana; fine sand, gypsiferous marl and occasional limestone.
Wasatch formation	Lower Eocene; Wyoming, south central and eastern Montana, southwestern North Dakota, western Colorado, Utah, and northwestern New Mexico; sands and clay, 0 to 5,000 ft +.

ated with extensive alteration in major faults occurring in granites, gneisses, and poorly-cemented sandstones under stress conditions commonly associated with tunneling projects, but it may be of concern in very deep, open excavations. The swelling acts primarily on sidewalls as a type of slow continuous plastic deformation under a constant load. Problems caused by this swell mechanism are usually of more concern where close tolerances and long-term stability are critical.

12-10. FROST ACTION.

Freezing can induce swelling or heaving of rock in excavations by the expansion of water within the rock mass. Although pore water freezing in porous rocks may be of some concern, the principal concern is freezing water in joints, bedding planes, and other openings in the rock. Since many of these discontinuities may have been relatively tight prior to freezing, a spalling effect from frost may induce a

nonrecoverable bulking of the rock and reduction in strength of the rock mass in addition to the temporary uplift by freezing. Preventive measures can include limiting excavation of final grades to warmer seasons, moisture controls or barriers, and layers of soil or insulation blankets in areas of special concern.

12-11. DESIGN CONSIDERATIONS.

If rock in an excavation is found to have a swelling potential, it may not be a serious concern unless structures are to be placed on the rock surface. With structures, swell and differential swell must then be considered and preventive techniques used. Some foundation design techniques for handling swell problems are summarized in Table 12-6.

SECTION III. SOIL-ROCK CONTACTS

12-12. GENERAL.

Some of the most difficult excavation problems occur in rock that has been severely weathered or altered. While it is generally assumed that bedrock will be easy to locate and identify, the assumption may not always be correct. In some cases, weathering can form a residual soil that grades into unweathered bedrock, with several rock-like soil or soil-like rock transitions in between. These residual soils, saprolites, and weathered rocks require special consideration, since they may have characteristics of both rock and soil that affect rock excavations and foundation performance.

TABLE 12-4. Swell Potential and Atterberg Limits

Index property (1)	Swelling Potential		
	Low (2)	Medium (3)	High (4)
Liquid limit	30-40	40-55	55-90
Plastic limit	15-20	20-30	30-60
Shrinkage limit[a]	35-25	25-14	14-8
Free swell[b]	20-40	40-70	70-180

[a]Poor correlation to swelling properties.
[b]Described by Katzir and David (1986).

TABLE 12-5. Summary of Swell Potential Tests

Test method (1)	Test procedure summary (2)	Remarks (3)
Free-swell test	This specimen is over-dried, granulated, and placed in a test tube. Water is added and the amount of volume expansion is recorded.	The rock structure is destroyed and the grain sizes are reduced.
Calculated-pressure test	An intact specimen is immersed in kerosene or mercury to determine its initial volume. The specimen is then placed in water and allowed to swell. If the specimen remains intact, the new volume can be determined by again immersing the specimen in mercury. Otherwise, the swell is recorded as the cange in volume of the water-specimen system.	The loading and confining pressures are not representative of in-situ conditions.
Unconfined swell test	The specimen is placed in a container and ames dials are set to one or more axes of the specimen. Water is added and the axial expansion is recorded.	The loading and confining pressures are not representative of in-situ conditions.
Nominal load test	A specimen is inserted in a consolidometer, a nominal seating load is applied (generally 200 psf or 0.10 kg/cm^2), and water added. The volume change is recorded by an ames dial.	The loading and confining pressures are not representative of in-situ conditions.
Calculated-pressure test	The specimen is placed in a consolidometer and subjected to a calculated overburden pressure. Free access to water is then permitted and the volume expansion recorded. Modifications of this test included rebounding the specimen to the original void ratio.	
Constant-volume test	The specimen is inserted in a consolidometer and a seating load applied. Water is added, and pressure on the specimen increased such that the total volume change of the specimen is zero. The final pressure is taken as the "swell pressure."	Under in-situ conditions, the volume may change resulting in a reduced final pressure.
Double-deadmeter test	Two similar specimens are placed in separate consolidometers. One specimen is subjected to calculated overburden pressures and the deformation recorded. The other specimen is allowed free access to water, is permitted to swell, and then is subjected to overburden pressure. The difference in deformations or strain of the two specimens at the overburden pressure is considered the potential swell of the material.	Results are more typical of field conditions.
Triaxial test	A specimen is consolidated to the in-situ pressure as evaluated by stress measurements of statistical analyses. Generally, the horizontal stress is greater than the applied vertical stress. The specimen is then allowed free access to water and the swell recorded.	This test is typical of field conditions, and can simulate high horizontal stress field.

TABLE 12-6. Design Techniques and Methods of Treatment for Swelling Rocks
(after Linder 1976)

(1) *Waterproofing below and around foundations*: Though successful in preventing drainage into the strata directly below the foundation, this method does not consider evaporation. The technique is best employed to prevent desiccation of strata during construction.

(2) *Rigid-box design*: The design of the foundation into separate reinforced concrete units or boxes that can withstand predicted stresses and deformations is a feasible yet expensive solution to swell.

(3) *Saturation and control*: Saturation of swell-susceptible strata before construction by ponding will help reduce swell after construction. However, if the water content is not maintained additional settlement will be experienced during the life of the structure. Also seasonal fluctuation of the availability of water may cause the structure to rise and fall periodically.

(4) *High loading points*: As swell is a function of both deformation and pressure, it was reasoned that foundations with high unit loadings should experience less swell. However, such foundations have met numerous problems including uplift on footings. Such foundations also influence only a small volume below the footing and swell may be experienced due to swell of deeper strata.

(5) *Replacement of the stratum*: A drastic, expensive, yet totally effective procedure for near-surface strata.

(6) *Piers*: The concept of placing the base of the foundation below swell-susceptible strata or where water content changes are expected to be minimal has also been employed with varying success. Problems such as side friction and water changes induced by construction must be considered.

(7) *Flexible construction*: For light structures, the division of the structure into units which can move independently of each other can be a practical solution. Differential heave between units will cause no stress to the structure and minor repair work will assure continuing service.

(8) *Raised construction*: A little-used alternative is to place the structure on a pile system raised above the surface. This would allow normal air circulation and evaporation below the structure and, if drainage is properly designed, should cause minimal disturbance to the water content of swell-susceptible strata.

12-13. WEATHERING PROFILES.

Chemical weathering is the primary cause of gradational soil-rock contacts, with the most prominent cases occurring in warm, humid climates. The result can be irregular or pinnacled rock covered by gradational materials composed of seamy, blocky rock, saprolite, and soil. The preferred case of an abrupt contact between soil and unweathered rock is not usually what is found. General descriptions of the zones in typical profiles for igneous and metamorphic rocks are given in Table 12-7. There are similarities in the development of these profiles. Weathering tends to dissolve the most soluble materials and alter the least stable minerals first, following rock mass discontinuities such as faults, joints, bedding planes, and foliations. The unweathered rock surface may be highly irregular due to solution and alteration along these openings. Engineering design and excavation considerations are dependent upon specific weathering profiles developed in certain rock types. These profiles include: massive igneous, extrusive igneous, metamorphic, carbonate, and shale.

A. Massive Intrusive Igneous Profiles. Rocks typical of massive igneous profiles include granites and other igneous rocks with relatively homogenous, isotropic texture. Since this type of rock has few or no bedding planes, foliations, or concentrations of minerals relatively susceptible to weathering, the existing joints, faults, and shear zones control the development of weathering. Stress-relief slabbing or sheeting joints subparallel to the ground surface also provide a path for chemical weathering, as shown in Figure 12-4. Saprolite (Zone IC in Figure 12-4) in this type of profile may retain the texture and orientation of the parent rock. Relict joints may still act as sliding-failure planes or preferred paths for ground water flow, so some of the parent rock's properties still apply to this material. The transition (Zone IIA) has the same slide failure and ground water concerns as with the saprolite, but the element of corestones becomes an additional concern. These are the hard, partially weathered spheroidal centers of blocks that can range from soft to relatively hard and from small size to relatively large. The transition zone may require a modification of the excavation methods used, from purely mechanical soil excavation methods to the occasional use of explosives or hand-breaking. Corestones in this type of profile are generally

TABLE 12-7. Description of a Typical Weathering Profile

	Zone (1)	Description (2)	RQD[a] (NX core, percent) (3)	Percent core recovery[b] (NX core) (4)	Relative permeability (5)	Relative strength (6)
I Residual soil	1A-A horizon	- Top soil, roots, organic material zone of leaching and eluviation may be porous	—	0	Medium to high	Low to medium
	1B-B horizon	- Characteristically clay-enriched, also accumulations of Fe, A1 and Si, hence may be cemented - No relict structures present	—	0	Low	Commonly low (high if cemented)
	1C-C horizon	- Relict rock structures retained - Silty grading to sandy material - Less than 10% core stones - Often micaceous	0 or not applicable	Generally 0-10%	Medium	Low to medium (relict structures very significant
II weathered rock	IIA-transition (from residual soil or saprolite to partly weathered rock)	- Highly variable, soil-like to rocklike - Fines commonly fine to coarse sand (USS) - 10 to 90% corestones - Spheroidal weathering common	Variable, generally 0-50	Variable, generally 10-90%	High (water losses common)	Medium to low where waste structures and relict structures are present
	IIB-partly weathered rock	- Rock-like, soft to hard rock - Joints stained to altered - Some alteration of feldspars and micas	Generally 50-75%	Generally >90%	Medium to high	Medium to high[b]
III unweathered rock		- No iron stains to trace long joints - No weathering of feld and micas	>75% (generally >90%)	Generally 199% 100%	Low to medium	Very high[b]

[a]The descriptions provide the only reliable means of distinguishing the zones.
[b]Considering only intact rock masses with no adversely oriented geologic structure.

spheroidal, which can cause difficulties in excavation and removal, if they are relatively large.

B. Extrusive Igneous Profiles. Extrusive rocks such as basalt develop profiles and conditions similar to those found in massive igneous rocks. However, certain structural features common in basalts and tuffs make conditions extremely variable in some areas. For example, lava flow tubes and vesicular basalt may increase the weathering path in some zones. The nature of flow deposits may make rock conditions in excavations difficult to predict since there may be buried soil profiles and interbedded ash falls or tuffs that are more permeable than adjacent basalts. These complex permeable zones can increase weathering and store water under relatively high pressures. Also, soils in the upper horizons may have unpredictable engineering characteristics due to unusual clay minerals present from the weathering of highly ferromagnesian parent materials.

C. Metamorphic Profiles. Since the structure or texture of metamorphic rocks can range from schistose to nearly massive gneissic, the weathering profiles can vary greatly, as illustrated in Figure 12-5. Foliations in the rocks and changes of the lithology enhance the variability that can be found in

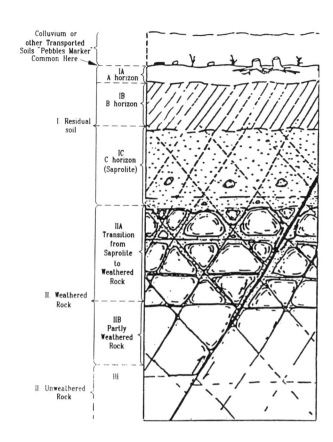

FIG. 12-4. Typical Weathering Profile for Intrusive Igneous Rocks (from Deere and Patton 1971)

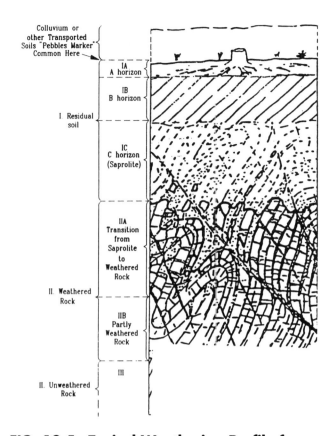

FIG. 12-5. Typical Weathering Profile for Metamorphic Rocks (from Deere and Patton 1971)

the weathering profiles in metamorphics. The results are differences in the depths of weathering profiles developed over each lithology, in some cases up to 50 m of difference vertically in just a few feet horizontally (Deere and Patton 1971). Intrusive dikes commonly found in metamorphic terrains may either be more or less resistant to weathering than the surrounding rock, forming either ridges or very deep weathering profiles. Problems in this type of profile include slide instability along relict foliation planes, highly variable depth to unweathered bedrock, and potentially high-pressure ground water storage in faults or behind intrusive dikes.

D. Carbonate Profiles. Carbonate rock weathering was previously discussed in relation to karst development within the rock mass. The same weathering conditions may affect the surface of the rock. Carbonate rocks develop into a profile, as illustrated in Figure 12-6, with sharp contacts between soils and weathered rock, unlike igneous, and metamorphic profiles. Occasionally, carbonate rocks may have chert, sand, or clay that form saprolite and retain a relict structure upon weather-

ing. In most cases, however, the carbonates are removed and the remaining insoluble residue, typically a dark red clayey "terra rosa," lies directly upon weathered rock. A jagged, pinnacled rock surface may develop due to weathering along faults or near-vertical joints. Troughs between the peaks may contain soft, saturated clays called "pockets of decalcification." Construction problems may include clayey seams, soft clays, rough bedrock surface, unstable collapse residuum, and rock cavities.

E. Shale Profile. Shale weathering profiles also develop primarily along joints and fissures, but the weathering profile is generally thinner and the transition from soil to unweathered rock tends to be more gradual. Shale is generally composed of minerals that are the weathering by-products of other rocks, so under a new weathering environment they are not affected to the extent other rocks are. Mechanical weathering mechanisms, such as drying and rewetting, freeze/thaw cycles, and stress relief play a more important role in the development of a shale weathering profile, so increased fracturing is the characteristic of increasingly weathered shale.

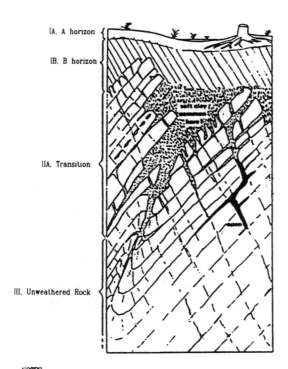

IA, A horizon

IB, B horizon

IIA, Transition

III, Unweathered Rock

NOTES:
1) Very impure (sandy or silty) carbonates may develope a saprolite, IC zone.
2) A partly weathered chalky limestone, IIB zone is sometimes present.

FIG. 12-6. Typical Weathering Profile for Carbonate Rocks (from Deere and Patton 1971)

Interbedded sandstones tend to make the weathering patterns and overall stability problems more complex. For engineering design purposes, the handling of shale excavations grades from rock mechanics into soil mechanics, where most weathered shales can be treated as consolidated clays.

12-14. DESIGN CONSIDERATIONS IN WEATHERING PROFILES.

During subsurface investigations, saprolites most likely are classified as soils, since the samples recovered by subsurface drilling programs frequently end up as a disaggregated, crumbly material with no apparent structure. The sampling technique frequently destroys the interparticle bonding and gives the designer a poor idea of the actual conditions. Care should be taken during sampling to determine if saprolites and relict structures exist if they will be exposed in rock excavations. Trenching provides a better picture of the weathering profile in critical areas.

A. Saprolites. Since relict discontinuities may exist in saprolite zones, sliding or toppling of weak blocks may be difficult to evaluate in stability analyses. In some cases, studies using key-block theory (Goodman and Shi 1985) may be applicable to saprolites. The discontinuities may also be the principal permeability path for ground water in saprolites, and water pressure in relict joints may play a substantial part in excavation stability. For design purposes, saprolites should be considered a weak, blocky, seamy rock in which discontinuities govern the behavior. For excavation purposes, saprolites may be treated as a firm soil, requiring standard soil excavation techniques.

B. Transition Materials. Below the saprolites in the weathering profile, the nature of the materials is more difficult to determine. The materials may act as a soil matrix with rock fragments of lesser importance, a rock mass with soil-like, compressible seams, or some intermediate material. The primary concern is the thickness compressibility or stability of the soil-like material between core-blocks or in seams, which governs the behavior of the material to a larger degree than the more easily recovered competent rock. In addition, rock in these zones may have an irregular surface, but may be adequate for load bearing. These conditions may require removal of all pinnacles to a prescribed suitable depth, or cleaning out of the crevices and backfilling with dental concrete. Lightly loaded footings on seamy rock may be adequate if the footings are expanded to prevent eccentric loading on individual blocks. If settlements are anticipated to be excessive using these techniques, drilled piers extending to competent rock at depth may be an economic alternative.

APPENDIX A

REFERENCES

A-1. REQUIRED PUBLICATIONS

TM 5-232. "Elements of Surveying, Headquarters, Department of the Army."

TM 5-235. "Special Surveys, Headquarters, Department of the Army."

TM 5-818-1/AFM 88-3 (Chapter 7). "Procedures for Foundation Design of Buildings and Other Structures."

ER 1110-1-1801. "Construction Foundation Report."

ER 1110-1-1802. "Provision for Spacers to Show Voids and Core Losses in Core Samples and Requirements for Photographic Record of Cores."

ER 1110-2-112. "Required Visits to Construction Sites by Design Personnel."

ER 1110-2-1806. "Earthquake Design and Analysis for Corps of Engineers Projects."

EP 1110-1-10. "Borehole Viewing Systems."

EM 385-1-1. "Safety and Health Requirements Manual."

EM 1110-1-1802. "Geophysical Exploration."

EM 1110-1-1804. "Geotechnical Investigations."

EM 1110-1-1904. "Settlement Analysis."

EM 1110-1-2907. "Rock Reinforcement."

EM 1110-2-1902. "Stability of Earth and Rockfill Dams."

EM 1110-2-1907. "Soil Sampling."

EM 1110-2-1908 (Part 1 of 2). "Instrumentation of Earth and Rockfill Dams (Ground Water and Pore Pressure Observations)."

EM 1110-2-1908 (Part 2 of 2). "Instrumentation of Earth and Rockfill Dams (Earth Movement and Pressure Measuring Devices)."

EM 1110-2-2200. "Gravity Dam Design."

EM 1110-2-2502. "Retaining and Flood Walls."

EM 1110-2-3504. "Chemical Grouting."

EM 1110-2-3506. "Grouting Technology."

EM 1110-2-3800. "Systematic Drilling and Blasting for Surface Excavations."

EM 1110-2-4300. "Instrumentation for Concrete Structures."

A-2. RELATED PUBLICATIONS.

American Society for Testing and Materials Standard Methods of Test D653. "Standard Terms and Symbols Relating to Soil and Rock." ASTM, 1916 Race St., Philadelphia, PA 19103.

Barton 1974. Barton, N. 1974. "A Review of the Shear Strength of Filled Discontinuities," Norwegian Geotechnical Institute, *NR 105*, 1–30.

Barton 1983. Barton, N. 1983. "Application of Q-System and Index Tests to Estimate Shear Strength and Deformability of Rock Masses." *Proceedings, International Symposium on Engineering Geology and Underground Construction,* Laboratorio Nacional de Engenharia Civil, Lisbon, Portugal, Vol. II, II.51–II.70.

Barton, Lien, and Lunde 1974. Barton, M., Lien, R., and Lunde, J. 1974. "Engineering Classification of Rock Masses for the Design of Tunnel Support." *Rock Mechanics,* Vol. 6, No. 4, 183–236.

Bieniawski 1973. Bieniawski, Z. T. 1973. "Engineering Classification of Jointed Rock Masses." *Transactions of the South African Institution of Civil Engineers,* Vol. 15, No. 12, 335–344.

Bieniawski 1978. Bieniawski, Z. T. 1978. "Determining Rock Mass Deformability: Experience from Case Histories." *International Journal of Rock Mechanics and Mining Sciences,* Vol. 15, 237–248.

Bieniawski 1979. Bieniawski, Z. T. 1979. "Tunnel Design by Rock Mass Classifications." *Technical Report GL-79-19,* U.S. Army Engineer Waterways Experiment Station, Vicksburg, MS.

Bishnoi 1968. Bishnoi, B. W. 1968. "Bearing Capacity of Jointed Rock," PhD thesis, Georgia Institute of Technology.

Bishop 1955. Bishop, A. W. 1955. "The Use of the Slip Circle in the Stability Analysis of Earth Slopes," *Geotechnique,* Vol. 5, 7–17.

Canada Centre for Mineral and Energy Technology 1977a. Canada Centre for Mineral and Energy Technology. 1977. *Pit Slope Man-*

ual. *CAMMET Report 77-15*, Chapter 5; "Design, Minerals Research Program." Mining Research Laboratories, available from Printing and Publishing Supply and Services, Canada, Ottawa, Canada.

Canada Centre for Mineral and Energy Technology 1977b. Canada Centre for Mineral and Energy Technology. 1977. *Pit Slope Manual. CAMMET Report 77-15*, Chapter 8; "Monitoring, Minerals Research Program." Mining Research Laboratories, available from Printing and Publishing Supply and Services, Canada, Ottawa, Canada.

Chan and Einstein 1981. Chan, H. C., and Einstein, H. H. 1981. "Approach to Complete Limit Equilibrium Analysis for Rock Wedges—The Method of Artificial Supports." *Rock Mechanics*, Vol.14, No. 2.

Cundall 1980. Cundall, P. A. 1980. "UDEC—A Generalized Distinct Element Program for Modeling Jointed Rock." European Research Office, U.S. Army, AD A087 610.

Deere 1964. Deere, D. U. 1964. "Technical Description of Rock Cores for Engineering Purposes." *Rock Mechanics and Engineering Geology*, Vol. 1, No. 1, 17–22.

Deere et al. 1967. Deere, D. U. et al. 1967. "Design of Surface and Near-Surface Construction in Rock." *Proceedings of the Eighth Symposium on Rock Mechanics* (ed. Fairhurst, C.), American Institute of Mining, Metallurgical and Petroleum Engineers, 237–302.

Deere and Deere 1989. Deere, D. U., and Deere, D. W. 1989. "Rock Quality Designation (RQD) After Twenty Years," *Technical Report GL-89-1*, U.S. Army Engineer Waterways Experiment Station, Vicksburg, MS.

Deere and Patton 1971. Deere, D. U., and Patton, F. D. 1971. "Slope Stability in Residual Soils." *Fourth Panamerican Conference on Soil Mechanics and Foundation Engineering*, American Society of Civil Engineers, 87–170.

Deere, Merritt, and Coon 1969. Deere, D. U., Merritt, A. H., and Coon, R. F. 1969. "Engineering Classification of In-Situ Rock." *Technical Report No. AFWL-TR-67-144*, Kirtland Air Force Base, New Mexico, 280 pp. Available from the U.S. Department of Commerce, NTIS, Springfield, VA, Pub. No. AD. 848 798.

Dickinson 1988. Dickinson, R. M. 1988. "Review of Consolidation Grouting of Rock Masses and Methods for Evaluation." U.S. Army Engineer Waterways Experiment Station, Vicksburg, MS.

Dowding 1985. Dowding, C. H. 1985. *Blast Vibration Monitoring and Control*, Prentice-Hall, Englewood Cliffs, NJ.

Dunnicliff 1988. Dunnicliff, J. 1988. *Geotechnical Instrumentation for Monitoring Field Performance*, Wiley and Sons, NY.

DuPont de Nemours 1977. DuPont de Nemours (E. I.) and Co., Inc. 1977. *Blaster's Handbook*. Wilmington, DE.

Farmer 1983. Farmer, I. 1983. *Engineering Behavior of Rocks*. Chapman and Hall, NY.

Goodman 1976. Goodman, R. E. 1976. *Methods of Geological Engineering in Discontinuous Rock*. West Publishing Company, St. Paul, MN.

Goodman 1980. Goodman, R. E. 1980. *Methods of Geological Engineering in Discontinuous Rock*. Wiley and Sons, NY.

Goodman and Shi 1985. Goodman, R. E., and Shi, G. 1985. *Block Theory and its Application to Rock Engineering*. Prentice-Hall, Inc., Englewood Cliffs, NJ.

Hanna 1973. Hanna, T. H. 1973. *Foundation Instrumentation*. Trans Tech Publications, Cleveland, OH.

Hendron, Cording, and Aiyer 1980. Hendron, A. J., Cording, E. J., and Aiyer, A. K. 1980. "Analytical and Graphical Methods for the Analysis of Slopes in Rock Masses." *Technical Report GL-80-2*, U.S. Army Engineer Waterways Experiment Station, Vicksburg, MS.

Hobst and Zajic 1977. Hobst, L., and Zajic, J. 1977. "Anchoring in Rock," *Developments in Geotechnical Engineering 13*. Elsevier Scientific Publishing Company, NY.

Hoek and Bray 1974. Hoek, E., and Bray, J. W. 1974. "Rock Slope Engineering." The Institution of Mining and Metallurgy, London, England.

Hoek and Bray 1981. Hoek, E., and Bray, J. W. 1981. "Rock Slope Engineering." The Institution of Mining and Metallurgy, London, England.

Hoek and Brown 1980. Hoek, E., and Brown, E. T. 1980. "Empirical Strength Criterion for Rock Masses." *Journal of Geotechnical Engineering*, American Society of Civil Engineers, Vol.106, No. GT9, 1013–1035.

Janbu 1954. Janbu, N. 1954. "Application of Composite Slip Surfaces for Stability Analysis." *Proceeding of the European Conference on Stability of Earth Slopes*, Sweden, Vol. 3, 43–49.

Janbu 1973. Janbu, N. 1973. "Slope Stability Computations." *Embankment Dam Engineering, The Casagrande Volume*. Wiley and Sons, NY, 47–86.

Janbu, Bjerrum, and Kjaernsli 1956.
Janbu, N., Bjerrum, L., and Kjaernsli, B. 1956. "Veiledning Ved Losing av Fandamentering-soppgaver" (Soil Mechanics Applied to Some Engineering Problems), in Norwegian with English Summary, *Norwegian Geotechnical Institute Publication No. 16*, Oslo.

Johnson 1989. Johnson, L. D. 1989. "Design and Construction of Mat Foundations," *Miscellaneous Paper GL-89-27*, U.S.Army Engineer Waterways Experiment Station, Vicksburg, MS.

Katzir and David 1968. Katzir, M., and David, P. 1968. "Foundations in Expansive Marls." *Proceedings, 2nd International Research in Engineering Conference on Expansive Clay Soils*, University of Texas.

Kovari and Fritz 1989. Kovari, K., and Fritz, P. 1989. "Re-evaluation of the Sliding Stability of Concrete Structures on Rock with Emphasis on European Experience," *Technical Report REMR-GT-12*, U.S. Army Engineer Waterways Experiment Station, Vicksburg, MS.

Kulhawy and Goodman 1980. Kulhawy, F. H., and Goodman, R. E. 1980. "Design of Foundations on Discontinuous Rock." *Proceedings of the International Conference on Structural Foundations on Rock*, International Society for Rock Mechanics, Vol. I, 9–220.

Lama and Vutukuri 1978. Lama, R. D., and Vutukuri, V. S. 1978. "Handbook on Mechanical Properties of Rocks." *Testing Techniques and Results—Vol. III*, No. 2, Trans Tech Publications (International Standard Book Number 0-87849-022-1, Clausthal, Germany).

Lambe and Whitman 1969. Lambe, T. W., and Whitman, R. V. 1969. *Soil Mechanics*. Wiley and Sons, NY.

Lauffer 1958. Lauffer, H. 1958. "Gebirgsklassifizierung fur den Stollenbau." *Geologie and Bauwesen*, Vol. 24, No. 1, 46–51.

Linder 1976. Linder, E. 1976. "Swelling Rock: A Review." *Rock Engineering for Foundations and Slopes*, American Society of Civil Engineers Specialty Conference, Boulder, CO, Vol. 1, 141–181.

Littlejohn 1977. Littlejohn, G. S. 1977. "Rock Anchors: State-of-the-Art." *Ground Engineering*, Foundation Publications Ltd., Essex, England.

Littlejohn and Bruce 1975. Littlejohn, G.S. and Bruce, D.A. 1975. "Rock Anchors: State-of-the-Art. Part I. Design." *Ground Engineering*, Foundation Publications Ltd., Essex, England.

Morgenstern and Price 1965. Morgenstern, N. R., and Price, V. E. 1965. "The Analysis of the Stability of General Slip Surfaces." *Geotechnique*, Vol. 15, 79–93.

Murphy 1985. Murphy, William L. 1985. "Geotechnical Descriptions of Rock and Rock Masses," *Technical Report GL-85-3*, U.S.Army Engineer Waterways Experiment Station, Vicksburg, MS.

Naval Facilities Engineering Command 1982. Naval Facilities Engineering Command. 1982. "Soil Mechanics Design Manual 7.1. NAVFAC DM-7.1." Department of the Navy, Alexandria, VA.

Nicholson 1983a. Nicholson, G. A. 1983a. "Design of Gravity Dams and Rock Foundations: Sliding Stability Assessment by Limit Equilibrium and Selection of Shear Strength Parameters." *Technical Report GL-83-13*, U.S. Army Engineer Waterways Experiment Station, Vicksburg, MS.

Nicholson 1983b. Nicholson, G. A. 1983b. "In-Situ and Laboratory Shear Devices for Rock: A Comparison." *Technical Report GL-83-14*, U.S. Army Engineer Waterways Experiment Station, Vicksburg, MS.

Peck, Hanson, and Thornburn 1974. Peck, R. B., Hanson, W. E., and Thornburn, T. H. 1974. *Foundation Engineering*, 2nd ed., Wiley and Sons, NY.

Poulos and Davis 1974. Poulos, H. G., and Davis, E. H. 1974. *Elastic Solutions for Soil and Rock Mechanics*. Wiley and Sons, NY.

Pratt et al. 1972. Pratt, H. R , et al. 1972. "The Effect of Specimen Size on the Mechanical Properties of Unjointed Diorite." *International Journal of Rock Mechanics and Mining Sciences and Geomechanics Abstracts*, Vol. 9, No. 4, 513–529.

Priest 1985. Priest, S. D. 1985. *Hemispherical Projection Methods in Rock Mechanics*. George Allen and Unwin (Publishers), Ltd., London England.

Rock Testing Handbook 1990. *Rock Testing Handbook*. 1990. Available from Technical Information Center, U.S. Army Engineer Waterways Experiment Station, 3909 Halls Ferry Road, Vicksburg, MS.

Serafim and Pereira 1983. Serafim, J. L., and Pereira, J. P. 1983. "Considerations of the Geomechanics Classification of Bieniawski." *Proceedings, International Symposium on Engineering Geology and Underground Construction*, LNEC, Lisbon, Portugal, Vol. 1, II.33–II.42.

Sarma 1979. Sarma, S. K. 1979. "Stability Analysis of Embankments and Slopes." *Journal of the Geotechnical Engineering Division*, American Society of Civil Engineers, ASCE, Vol. 105, No. GT 12, 1511–1524.

Sowers 1979. Sowers, G. F. 1979. *Introductory Soil Mechanics and Foundations: Geotechnical Engineering*, 4th ed., McMillan, NY.

Templeton 1984. Templeton, A. E. 1984. "User's Guide: Computer Program for Determining Induced Stresses and Consolidation Settlement (CSETT)." *Instruction Report K-84-7*, U.S. Army Engineer District, Vicksburg, MS.

APPENDIX B

ROCK MASS CLASSIFICATION TABLES

CLASSIFICATION INPUT DATA WORKSHEET: GEOMECHANICS CLASSIFICATION OF ROCK MASSES

Name of project: _____
Site of survey: _____
Conducted by: _____
Date: _____

STRUCTURAL REGION
No. _____
Sta. _____
Sta. _____
Sta. _____
Sta. _____

ROCK TYPE AND ORIGIN

DRILL CORE QUALITY R.Q.D.

Excellent:	90-100% ....
Good:	75-90% ....
Fair:	50-75% ....
Poor:	25-50% ....
Very poor:	<25% ....
Range %:	

WALL ROCK OF DISCONTINUITIES

Unweathered	....
Slightly weathered	....
Moderately weathered	....
Highly weathered	....
Completely weathered	....

CONDITION OF DISCONTINUITIES

CONTINUITY

		Set 1	Set 2	Set 3
Very low:	<3 ft	....	....	....
Low:	3-10 ft	....	....	....
Medium:	10-30 ft	....	....	....
High:	>30 ft	....	....	....

SEPARATION

		Set 1	Set 2	Set 3
Tight joints:	<0.01	....	....	....
Moderately open joints:	0.01-0.1 in.	....	....	....
Open joints:	0.1-0.5 in.	....	....	....
Range in.:				

ROUGHNESS

	Set 1	Set 2	Set 3
Very rough surfaces:	....	....	....
Rough surfaces:	....	....	....
Slightly rough surfaces:	....	....	....
Smooth surfaces:	....	....	....
Slickensided surfaces:	....	....	....

FILLING (GOUGE)

	Set 1	Set 2	Set 3
Type:	....	....	....
Thickness:	....	....	....
Consistency:	....	....	....

GROUNDWATER

INFLOW per 1,000 ft .... gal/min
of tunnel length
or
GENERAL CONDITIONS (completely dry, damp, wet, dripping or flowing under low, medium or high pressure):

STRENGTH OF INTACT ROCK MATERIAL

	Uniaxial compressive strength, psi	Set 1	Set 2	Set 3
Very high:	Over 32,000	....	....	....
High:	16,000 - 32,000	....	....	....
Medium:	8,000 - 16,000	....	....	....
Low:	4,000 - 8,000	....	....	....
Very low:	150 - 4,000	....	....	....

MAJOR FAULTS OR FOLDS

Describe major faults and folds specifying their locality, nature, and orientation.

SPACING OF DISCONTINUITIES

		Set 1	Set 2	Set 3
Very wide:	Over 10 ft	....	....	....
Wide:	3-10 ft	....	....	....
Moderate:	1-3 ft	....	....	....
Close:	2 in. - 1 ft	....	....	....
Very close:	<2 in.	....	....	....
Range:				

GENERAL REMARKS AND ADDITIONAL DATA

The geologist should supply any further information which he considers relevant.

STRIKE AND DIP ORIENTATIONS

			DIRECTION
Set 1	Strike:	(from .... to ....)	Dip: ....
Set 2	Strike:	(from .... to ....)	Dip: ....
Set 3	Strike:	(from .... to ....)	Dip: ....

TABLE B-1. Input Data Sheet for Geomechanics (RMR) Classification System

TABLE B-2. Geomechanics Classification of Jointed Rock Masses

PARAMETER			RANGES OF VALUES					
(a) Classification Parameters and their ratings								
1	Strength of intact rock material	Point-load strength index	>10 MPa	4–10 MPa	2–4 MPa	1–2 MPa	For this low range uniaxial compressive test is preferred		
		Uniaxial compressive strength	>250 MPa	100–250 MPa	50–100 MPa	25–50 MPa	5–25 MPa	1–5 MPa	<1 MPa
		Rating	15	12	7	4	2	1	0
2	Drill core quality RQD		90–100%	75–90%	50–75%	25–50%	<25%		
	Rating		20	17	13	8	3		
3	Spacing of discontinuities		>2 m	0.6–2 m	200–600 mm	60–200 mm	<60 mm		
	Rating		20	15	10	8	5		
4	Condition of discontinuities		Very rough surfaces Not continuous No separation Unweathered wall rock	Slightly rough surfaces Separation < 1 mm Slightly weathered walls	Slightly rough surfaces Separation <1 mm Highly weathered walls	Slickensided surfaces OR Gouge < 5 mm thick OR Separation 1–5 mm Continuous	Soft gauge > 5 mm thick OR Separation > 5 mm Continuous		
	Rating		30	25	20	10	0		
5	Ground water	Inflow per 10 m tunnel length	None	< 10 L/min	10–25 L/min	25–125 L/min	> 125		
			OR	OR	OR	OR	OR		
		Ratio *Joint water pressure / major principal stress*	0	0.0–0.1	0.1–0.2	0.2–0.5	>0.5		
			OR	OR	OR	OR	OR		
		General conditions	Completely dry	Damp	Wet	Dripping	Flowing		
	Rating		15	10	7	4	0		
(b) Rating Adjustment for Joint Orientations								
Strike and dip orientations of joints			Very favorable	Favorable	Fair	Unfavorable	Very unfavorable		
Ratings	Tunnels		0	−2	−5	−10	−12		
	Foundations		0	−2	−7	−15	−25		
	Slopes		0	−5	−25	−50	−80		
(c) Rock Mass Classes Determined from Total Ratings								
Rating			100←81	80←61	60←41	40←21	<20		
Class number			I	II	III	IV	V		
Description			Very good rock	Good rock	Fair rock	Poor rock	Very poor rock		
(d) Meaning of Rock Mass Classes								
Class number			I	II	III	IV	V		
Average stand-up time			10 years for 15 m span	6 months for 8 m span	1 week for 5 m span	10 hours for 2.5 m span	30 minutes for 1 m span		
Cohesion of rock mass			>400 kPa	300–400 kPa	200–300 kPa	100–200 kPa	< 100 kPa		
Friction angle of rock mass			>45°	35–45°	25–35°	15–25°	<15°		

(a)

Dip 0° - 10°	Dip 10° - 30°		Dip 30° - 60°	Dip 60° - 90°
	Dip direction			
	Upstream	Downstream		
Very favorable	Unfavorable	Fair	Favorable	Very unfavorable

(b)

Strike Perpendicular to Tunnel Axis			
Drive with Dip		Drive against Dip	
Dip 45°-90°	Dip 20°-45°	Dip 45°-90°	Dip 20°-45°
Very favorable	Favorable	Fair	Unfavorable

Strike Parallel to Tunnel Axis		Dip 0°-20° Irrespective of Strike
Dip 45°-90°	Dip 20°-45°	
Very unfavorable	Fair	Fair

TABLE B-3. Summary of Joint Orientation Adjustments for Dam Foundations and Tunnels: (a) Assessment of Joint Orientation Favorability on Stability of Dam Foundations; (b) Effect of Joint Strike and Dip Orientations in Tunneling

Rock Quality Designation (RQD)

Very poor.....................	0-25	**Note:**
Poor.....................	25-50	(i) Where RQD is reported or measured as ≤ 10 (including 0) a nominal value of 10 is used to evaluate Q in Eq. (1).
Fair.....................	50-75	
Good.....................	75-90	
Excellent.....................	90-100	(ii) RQD intervals of 5, i.e. 100, 95, 90 etc. are sufficiently accurate.

Joint Set Number (J_n)

Massive, no or few joints	0.5-1.0	**Note:**
One joint set............	2	(i) For intersections use $(3.0 \times J_n)$
One joint set plus random	3	
Two joint sets...........	4	(ii) For portals use $(2.0 \times J_n)$
Two joint sets plus random..................	6	
Three joint sets.........	9	
Three joint sets plus random..................	12	
Four or more joint sets, random, heavily jointed, "sugar cube", etc........	15	
Crushed rock, earthlike..	20	

Joint Roughness Number (J_r)

(a) Rock wall contact and (b) Rock wall contact before 10 cms shear		**Note:**
		(i) Add 1.0 if the mean spacing of the relevant joint set is greater than 3 m.
Discontinuous joints.....	4	
Rough or irregular, undulating..............	3	
Smooth, undulating.......	2	**Note:**
Slickensided, undulating	1.5	(ii) $J_r = 0.5$ can be used for planar slickensided joints having lineation, provided the lineations are favorably orientated.
Rough or irregular, planar..................	1.5	
Smooth, planar...........	1.0	
Slickensided, planar.....	0.5	(iii) Descriptions B to G refer to small scale features and intermediate scale features, in that order.
(c) No rock wall contact when sheared		
Zone containing clay minerals thick enough to prevent rock wall contact	1.0 (nominal)	
Sandy, gravelly or crushed zone thick enough to prevent rock wall contact..................	1.0 (nominal)	

(Continued)

TABLE B-4. Q-System: Description and Ratings for RQD, J_n, J_r, J_s, and J_w Parameters (from Barton, Lien, and Lunde 1974)

Joint Alteration Number

	(J_a)	ϕ_r (approx.)
(a) Rock wall contact		
A. Tightly healed, hard, nonsoftening, impermeable filling i.e. quartz or epidote..........................	0.75	(-)
B. Unaltered joint walls, surface staining only......................	1.0	(25°-35°)
C. Slightly altered joint walls. Non-softening mineral coatings, sandy particles, clay-free disintegrated rock etc..........................	2.0	(25°-30°)
D. Silty-, or sandy-clay coatings, small clay-fraction (non-softening)	3.0	(20°-25°)
E. Softening or low friction clay mineral coatings, i.e. kaolinite, mica. Also chlorite, talc, gypsum and graphite etc., and small quantities of swelling clays. (Discontinuous coatings, 1-2 mm or less in thickness)................	4.0	(8°-16°)
(b) Rock wall contact before 10 cms shear		
F. Sandy particles, clay-free disintegrated rock etc.............	4.0	(25°-30°)
G. Strongly over-consolidated, non-softening clay mineral fillings (Continuous, <5 mm in thickness)....	6.0	(16°-24°)
H. Medium or low over-consolidation, softening, clay mineral fillings. (continuous, <5 mm in thickness)...	8.0	(12°-16°)
J. Swelling clay fillings, i.e. montmorillonite (Continuous, <5 mm in thickness). Value of J_a depends on percent of swelling clay-size particles, and access to water etc......................	8.0-12.0	(6°-12°)
(c) No rock wall contact when sheared		
K., L., M. Zones or bands of disintegrated or crushed rock and clay (see G., H., J. for description of clay condition)........................	6.0, 8.0 or 8.0-12.0	(6°-24°)
N. Zones or bands of silty- or sandy clay, small clay fraction (nonsoftening)......................	5.0	
O., P., R. Thick, continuous zones or bands of clay (see G., H., J. for description of clay condition).....	10.0, 13.0 or 13.0-20.0	(6°-24°)

Note:

(i) Values of $(\phi)_r$ are intended as an approximate guide to the mineralogical properties of the alteration products, if present.

(Continued)

TABLE B-4. *(continued)*

Stress Reduction Factor

(SRF)

(a) Weakness zones intersecting excavation, which may cause loosening of rock mass when tunnel is excavated.

A. Multiple occurrences of weakness zones containing clay or chemically disintegrated rock, very loose surrounding rock (any depth)............ 10.0

B. Single weakness zones containing clay, or chemically disintegrated rock (depth of excavation $\leq$50 m).................................... 5.0

C. Single, weakness zones containing clay, or chemically disintegrated rock (depth of excavation >50 m).................................... 2.5

D. Multiple shear zones in competent rock (clay free), loose surrounding rock (any depth)...... 7.5

E. Single shear zones in competent rock (clay free) (depth of excavation $\leq$50 m)............... 5.0

F. Single shear zones in competent rock (clay free) (depth of excavation >50 m)............. 2.5

G. Loose open joints, heavily jointed or "sugar cube" etc. (any depth)........................ 5.0

(b) Competent rock, rock stress problems.

		σ_c/σ_1	σ_t/σ_1	
H.	Low stress, near surface..	>200	>13	2.5
J.	Medium stress.............	200-10	13-0.66	1.0
K.	High stress, very tight structure (Usually favorable to stability, may be unfavorable to wall stability)................	10-5	0.66-0.33	0.5-2.0
L.	Mild rock burst (massive rock).....................	5-2.5	0.33-0.16	5-10
M.	Heavy rock burst (massive rock).....................	<2.5	<0.16	10-20

(c) Squeezing rock; plastic flow of incompetent rock under the influence of high rock pressures.

N. Mild squeezing rock pressure................... 5-10

O. Heavy squeezing rock pressure................... 10-20

(d) Swelling rock; chemical swelling activity depending on presence of water

P. Mild swelling rock pressure.................... 5-10

R. Heavy swelling rock pressure................... 10-15

Note:

(i) Reduce these values of SRF by 25-50% if the relevant shear zones only influence but do not intersect the excavation.

(ii) For strongly anisotropic stress field (if measured): when $5 \leq \sigma_1/\sigma_3 \leq 10$, reduce σ_c and σ_t to 0.8 σ_c and 0.8 σ_t; when σ_1/σ_3 > 10, reduce σ_c and σ_t to 0.6 σ_c and 0.6 σ_t where: σ_c = unconfined compression strength, σ_t = tensile strength (point load), σ_1 and σ_3 = major and minor principal stresses.

(iii) Few case records available where depth of crown below surface is less than span width. Suggest SRF increase from 2.5 to 5 for such cases (see H).

Joint Water Reduction Factor

		(J_w)	Approx. water pressure (kg/cm^2)
A.	Dry excavations or minor inflow, i.e. 5 l/min. locally...	1.0	<1
B.	Medium inflow or pressure occasional outwash of joint fillings...............................	0.66	1.0-2.5
C.	Large inflow or high pressure in competent rock with unfilled joints...............................	0.5	2.5-10.0
D.	Large inflow or high pressure, considerable outwash of joint fillings.....................	0.33	2.5-10.0
E.	Exceptionally high inflow or water pressure at blasting, decaying with time...................	0.2-0.1	>10.0
F.	Exceptionally high inflow or water pressure continuing without noticeable decay...........	0.1-0.05	>10.0

Note:

(i) Factors C to F are crude estimates. Increase J_w if drainage measures are installed.

(ii) Special problems caused by ice formation are not considered.

TABLE B-4. *(continued)*

INDEX

125